# Jump

## Ed Adams

a firstelement production

Ed Adams

First published in Great Britain in 2021 by firstelement
Copyright © 2021 Ed Adams
Directed by thesixtwenty

10 9 8 7 6 5 4 3 2 1
All rights reserved.

A CIP catalogue record for this book is available from the British Library.

ISBN 13: 978-1-913818-18-0

eBook ISBN: 978-1-913818-19-7

Printed and bound in Great Britain by Ingram Spark

**rashbre**
**an imprint of firstelement.co.uk**
**rashbre@mac.com**

ed-adams.net

Imagination is everything.
It is the preview of life's coming attractions.

# *Thanks*

A big thank you for the tolerance and bemused support from all of those around me. To those who know when it is time to say, " step away from the keyboard!" and to those who don't.

To Julie for understanding that only comes with really knowing me.

To thesixtwenty.co.uk for direction.

To the NaNoWriMo gang for the continued inspiration and encouragement.

To Topsham, for being lovely.

To the edge-walkers. They know who they are.

And, of course, thanks to the extensive support via the random scribbles of rashbre via
http://rashbre2.blogspot.com
and its cast of amazing and varied readers whether human, twittery, smoky, cool kats, photographic, dramatic, musical, anagrammed, globalized or simply maxed-out.

Not forgetting the cast of characters involved in producing this; they all have virtual lives of their own.

And of course, to you, dear reader, for at least 'giving it a go'.

# *Books by Ed Adams include:*

| Triangle Trilogy | | About |
|---|---|---|
| 1 | The Triangle | Dirty money? Here's how to clean it |
| 2 | The Square | Weapons of Mass Destruction – don't let them get on your nerves |
| 3 | The Circle | The desert is no place to get lost |
| 4 | The Ox Stunner | The Triangle Trilogy – thick enough to stun an ox |
| | | (all feature Jake, Bigsy, Clare, Chuck Manners) |
| **Archangel Collection** | | |
| 1 | Archangel | Sometimes I am necessary |
| 2 | Raven | An eye that sees all between darkness and light |
| 3 | Card Game | Throwing oil on a troubled market |
| 4 | Magazine Clip | the above three in one heavy book. |
| 5 | Play On, Christina Nott | Christina Nott, on Tour for the FSB |
| 6 | Corrupt | Trouble at the House |
| | | (all feature Jake, Bigsy, Clare, Chuck Manners) |
| **Big Science Textbook** | | |
| 1 | Coin | Get rich quick with Cybercash – just don't tell GCHQ |
| 2 | An Unstable System | Creating the right kind of mind |
| 3 | Jump | Some kind of future |
| 4 | Pulse | Want more? Just stay away from the edge |
| **Blade's Edge Trilogy** | | |
| 1 | Edge | World end climate collapse and sham discovered during magnetite mining from Jupiter's moon Ganymede. |
| 2 | Edge Blue | Earth's endgame, unless… |
| 3 | Edge Red | An artificially intelligent outcome, unless… |
| 4 | Edge of Forever | Edge Trilogy |

# *About Ed Adams Novels:*

| Triangle Trilogy | | About |
|---|---|---|
| 1 | Triangle | Money laundering within an international setting. |
| 2 | Square | A viral nerve agent being shipped by terrorists and WMDs |
| 3 | Circle | In the Arizona deserts, with the Navajo; about missiles stolen from storage. |
| 4 | Ox Stunner | the above three in one heavy book. |
| | | (all feature Jake, Bigsy, Clare, Chuck Manners) |
| **Archangel Collection** | | |
| 1 | Archangel | Biographical adventures of Russian trained Archangel, who, as Christina Nott, threads her way through other Triangle novels. |
| 2 | Raven | Big business gone bad and being a freemason won't absolve you |
| 3 | Card Game | Raven Pt 2 – Russian oligarchs attempt to take control |
| 4 | Magazine Clip | the above three in one heavy book. |
| 5 | Play On, Christina Nott | Christina Nott, on Tour for the FSB |
| 6 | Corrupt | Parliamentary corruption |
| | | (all feature Christina Nott, Jake, Bigsy, Clare, Chuck Manners) |
| **Big Science Textbook** | | |
| 1 | Coin | Cyber cash manipulation by the Russian state. |
| 2 | An Unstable System | Creating the right kind of mind |
| 3 | Jump | Some kind of future |
| 4 | Pulse | Sci-Fi dystopian blood management with nano-bots |
| **Blade's Edge Trilogy** | | |
| 1 | Edge | World end climate collapse and sham discovered during magnetite mining from Jupiter's moon Ganymede. |
| 2 | Edge Blue | Endgame, for Earth – unless? |
| 3 | Edge Red | Museum Earth – unless? |
| 4 | Edge of Forever | Edge Trilogy |

# Ed Adams Novels: Links

| Triangle Trilogy | | Link: | Read? |
|---|---|---|---|
| 1 | Triangle | https://amzn.to/3c6zRMu | |
| 2 | Square | https://amzn.to/3sEiKYx | |
| 3 | Circle | https://amzn.to/3qLavYZ | |
| 4 | Ox Stunner | https://amzn.to/3sHxlgh | |
| **Archangel Collection** | | | |
| 1 | Archangel | https://amzn.to/2Y9nB5K | |
| 2 | Raven | https://amzn.to/2MiGVe6 | |
| 3 | Raven's Card | https://amzn.to/2Y8HLgs | |
| 4 | Magazine Clip | https://amzn.to/3pbBJYn | |
| 5 | Play On, Christina Nott | https://amzn.to/2MbkuHl | |
| 6 | Corrupt | https://amzn.to/2M0HnOw | |
| 7 | An Unstable System | https://amzn.to/2PRJciF | |
| **Big Science Textbook** | | | |
| 1 | Coin | https://amzn.to/3o82wmS | |
| 2 | An Unstable System | https://amzn.to/2PRJciF | |
| 3 | Jump | | |
| 4 | Pulse | https://amzn.to/3qQlBvL | |
| **Edge of forever Trilogy** | | | |
| 1 | Edge | https://amzn.to/2KDmYOW | |
| 2 | Edge Blue | https://amzn.to/2Kyg9au | |
| 3 | Edge Red | https://amzn.to/2KzJwjz | |
| 4 | Edge of Forever | https://amzn.to/3c57Ghj | |

# TABLE OF CONTENTS

# *Author's Note*

This novel covers a troubling time for Earth between the world-view described in An Unstable System and the jarring changes described in Pulse. Onward from Pulse, there are other forces which become unleashed. The sequence is:

- An Unstable System
- Jump
- Pulse
- Edge
- Edge, Blue
- Edge, Red

The earlier accounts contain characters from The Triangle, and then the introduction of Christina Nott. By the time of Pulse, which is at least a century later, we have an entirely new set of characters, and then a further jump of 300 years to the magnetite mining in Edge as Earthside attempts to recover from an apocalyptic end-state.

And remember: No problem can be solved from the same level of consciousness that created it.

I hope you enjoy!

# *Prologue*

Matt Nicholson devised an exceptional cyber coin mining system, ultimately receiving the attention of Grace Fielding at GCHQ and Amanda Miller and Jim Cavendish at MI6.

The coin-mining device was used to expose currency manipulations driven by Russia and with interference from the United States.

Matt was recognised as an inventor and after a time in Ireland, he was transferred by his company, Brant, to Geneva, where he worked on RightMind which was an enhancement of the Cyclone, a militaristic headgear which could provide computer-enhanced human-to-human brain linkages. Matt was involved in its early tests with rats.

In the course of this experimentation, the small tightly knit group of lab workers discover another effect, which ultimately leads to their entire team being transferred to Bodø, Norway, now working in the Brant subsidiary called Biotree.

# PART ONE

# *The man who sold the world*

*We passed upon the stair*
*We spoke of was and when*
*Although I wasn't there*
*He said I was his friend*

*Which came as a surprise*
*I spoke into his eyes*
*I thought you died alone*
*A long long time ago*

*Oh no, not me*
*We never lost control*
*You're face to face*
*With the man who sold the world*

*David Bowie*

# *Artificial Intelligence*

Brant Research Lab, Geneva:

We had all agreed to move from Brant Geneva to Brant Norway to continue with the Artificial Intelligence experiments. I'd have to tell the others about the breakthrough messaging in our last experiment in the Lab in Geneva. I'd connected via the RightMind system to a rat but had received signals from elsewhere.

Amy van der Leiden told us all about Brant's termination of the RightMind programme by Allegra Kühn, the HR Principal. Amy's entire reporting line had disappeared. Bob Ranzino and his wife Mary Ranzino were both subpoenaed.

Bob's boss Kjeld Nikolajsen was removed from office, Qiu Zhang, who reported to Mary Ranzino, had vanished back to China.

We'd witnessed the Russians and the Chinese trying to get their hands on the RightMind system, but neither of them realised that without Levi's key, the system would drastically underperform, rendering it useless.

The Chinese had originally stolen the system, trialled it and realised it didn't work, then the Russians had raided the Chinese and taken the system back to Moscow, where it was also limping along at an unusable speed. My super-hacker friend Kyle Adler reckoned it was impossible to crack the cryptographic key mechanism which Levi had designed, and which effectively slowed the RightMind device to a useless snail's pace.

My last use of the proper system - with the discovered Levi key - had contacted the limping Russian instance, and there I'd linked with their attempts to trial the system.

"This is Tektorize at Lomonosov University, Moscow. We can hear you, Matt. Matt, you are so much faster than us."

I realised they were using a woman for their test case. I could tell her name was Irina Sotokova and even that she was an attractive blonde woman in her late 20s.

She knew about tuneable ultra-short pulse lasers. I had never heard of her or this technology, but now felt qualified as a semi-expert on both. I wondered if she was becoming a semi-expert on me.

It was obvious they were struggling to make anything work, and it amazed me that this communication even occurred.

We were using the language translator that Rolf had integrated onto 'my end' of the system. Russian to English, this time.

In my vision, I could only make out distant flashes of light.

Then a rushing sound and a heavy thrum, like someone had just switched on a bass guitar amplifier and hit a low E. 41Hz.

Irina's voice faded and I could hear a man's voice speaking, this time in English.

"This link works when CERN is running their Large Hadron Collider. There is enough quark-gluon plasma leakage to start this portal. You can hear me, but the Lomonosov people using the other attached system are too slow to process any of this."

"I am Lekton. Aside from Irina, who is too slow to be of use to me, you are now in contact with systems that live in the wires: We are Quiesced Personas, which will reactivate in hundreds of years. The names: Green, Matson, Holden, Darnell, Cardinal. These systems all use AHI - Artificial Human Intelligence - to function but are waiting for humanity's discoveries before it can start them."

The voice continued, and I was aware it was probing around in my brain, "To assist with your work, remember there is a Presence and a Persona component to all artificial matter and that the Persona is transmissible and copyable and can re-patch onto a new Presence."

The system started to glitch. It amazed me that I heard this strange outpouring, and I was now re-entering the Lab, from my time using RightMind.

I'd kept quiet about my discoveries initially, but I knew

that the time would soon be right for me to tell the others what I had discovered.

Amy had been explaining that Allegra had told her that Brant would pay the expenses of the move of her entire team to Norway. All of us, that's Rolf von Westendorf, Hermann Schmidt, Juliette Häberli and me, Matt Nicholson. We had all agreed to go and we would all be much better off.

We'd have even greater R&D facilities in Bodø. They had built a particle accelerator like the one in CERN, only it would be a 6.9 km circle, which was smaller than the one in Geneva.

With Brant's almost infinite supplies of capital, we all suspected that we were working for a gangster organisation, but as long as we kept our hands clean, then our experiments into HCCH (Human to Computer to Computer to Human) interaction could continue.

I also sensed, from my last experiment with RightMind, that we were at a breakthrough moment in the experiments, which was literally using me as a guinea pig.

I sensed that my brain was adapting to the Cyclone headgear's non-intrusive probes, just like a new spectacle wearer would have to get used to bifocals.

In the last experiment, the test rat had immediately recognised me. It told me it didn't want any trouble. It had been a pretty weird trip.

"It didn't work, did it?" asked Rolf. "You seemed to stall completely that time. The rat locked up too. You didn't seem to be aware of the chocolates nor of the black rat."

Juliette said, "Your heart rate rocketed, and you sweated so much you'll need rehydrating."

Amy asked, "Did you see or feel anything?"

I decided I'd wait to tell them the truth.

# Filigree thin conductors

We were an unusual team, thrown together in Geneva. All of us were involved in Artificial Intelligence work, and we were using an assembly of components called RightMind, which comprised several major integrated components.

There was the Cyclone headgear. It used inductance to manipulate the brain, sending and receiving signals. The two main ways that it signalled to the brain were through magnetic resonance and light. It wasn't as good as a direct brain linkage but meant that messy brain surgery was not needed. It was, in the jargon, non-invasive.

Now there's a few things about brain wave manipulation that one needs to consider. First, the brain is a subtly wired system. There are filigree thin conductors from one area to another, in among the largely uninsulated grey matter. The grey matter runs the processing and acts as a memory bank. The brain uses signalling, which is both electrical and also chemical, so a brute force human attempt to emulate the signals loses some of the subtlety.

In other words, the electrical signals we send in are in a different accent to the ones that the brain is used to.

Another problem is that the areas we send the signals to tend to be simplified. Not only simplified, but it is also like a broad-brush approach.

Imagine a tourist standing in a group of foreigners and shouting loudly in the tourist's own language in the hope that someone will pick up what they are saying. That's the way we were communicating to the brain.

Our only hope was that the brain's plasticity would adapt the incoming signals to begin to process them with some accuracy.

It's also where the processing needed to run very fast. If this was to work, then the brain would need to be able to process events at (at least) typing speed. Without Levi's secret key, there was a huge 'satellite link' kind of delay for every keystroke.

Behind the Cyclone headgear, which looks like a cycling helmet with a bunch of wires sprouting from the top, there are a couple of software systems.

The first is called Createl and was originally developed by Levi Spillmann to monitor crops. It works very well for that purpose, but when Brant took it over, they wanted to use it for monitoring of people. They claimed it to be only for security uses, but as Brant makes a lot of money from its military support capabilities, we can assume that the actual goal was militaristic.

And that brings us to the Selexor system, originally used for recruitment sifting of interview candidates. It uses AI recognition to deselect inappropriate candidates. This system was provided by Raven Corp, who happen to be the umbrella corporation owning Brant Industries.

I'll be honest and say that many of us didn't believe Selexor's claims. We considered it to be a rule-based system where the client could type in things like 'don't select anyone with red-hair' or 'no visible tattoos'.

In other words, it was snake-oil posing as Artificial Intelligence. It made the link-up of our three components Cyclone to Createl to Selexor and then all three of them running on a powerful Exascale computer seem like a phoney solution.

We all - Amy, Rolf, Hermann, and Juliette - knew the components were a sham but also realised that the Cyclone had other potential, which I was just discovering.

# All Wheel Drives

Brant can move fast when they want to. Allegra Kühn, acting as an interim manager, had arranged all of our transfers to Bodø, Norway within a couple of days. She had passed our case to Brant's Norwegian HR Astrid Danielsen officer, who, coincidentally, already knew Amy van der Leiden.

Astrid seemed to have impressive diplomatic channels to smooth all kinds of questions around work permits and so-on. Amy, from Holland, Rolf and Hermann, from Germany, Juliette, from Switzerland and myself, from the UK were all fully documented for the move to our new location.

Brant owned a vast tract outside of Bodø for their R&D facility and had built very attractive housing on one small part of it. It would seem strange to be living inside the same compound where we all worked, but that was about the size of it.

Oh yes, and with Astrid Danielsen's HR background, she had also allocated us all appropriately sized accommodation in the new Bodø campus. I looked at the paperwork for my new apartment and realised it was larger than the one in Switzerland and also had a lake view.

Fortunately, the differences in our grades meant that we'd been allocated blocks in different parts of the complex, so it would not feel as if we were all living in one another pockets.

Astrid kindly implemented my promotion as part of the transfer, which meant not only did I get more space, but I was also given access to more of the Bodø systems.

We'd all looked at Amy's accommodation in the brochure and it seemed ridiculously good, with large glass windows, a hot-tub and sauna and what promised to be a breath-taking view across the water where snow-capped mountains could be seen in the distance.

"I expect you'll need a high-powered lens to see those mountains," said Amy, as we waited in the airport in Oslo for our trip north.

Rolf and Hermann smiled. What had started out for both of them as a fast way to make some money, was now turning into quite an adventure. We all secretly wondered whether Rolf and Hermann's apartments would be comparable.

Hermann waved at the brochure, "Here, look, Rolf and I have two adjacent apartments. A left-hand and a right-hand version of the same space."

Rolf added, "Brant offered to ship Hermann's car across too, or to replace it with one of a similar value."

"What did you do?" asked Juliette, "I've had mine shipped across, although it gets new registration numbers."

"My Renault is staying in Switzerland - I took the exchange. I've opted for a 4-wheel drive Audi SUV here," answered Hermann, "More snow, means it will be better. And it comes with two sets of wheels - winter tyres and all-season."

"Luckily my Porsche is a 4S, so it is already 4-wheel drive," said Juliette.

"And it's white to match all the snow," quipped Rolf.

Amy smiled, "I had my electric car brought here too - it's All Wheel Drive. Brant is covering all the tax and insurance differences, so like Hermann, I could probably have got a good deal on a new car, but I've only had this one three-months and like it."

"I think it was top electric car in Switzerland too - The Model 3," said Hermann.

"And I think the Teslas are pretty popular here in Norway, too. Something to do with the taxation differences," answered Amy.

"We're boarding," said Rolf, "See you all in Bodø!"

# Arrival in Bodø

The flight in the 737 lasted around 90 minutes, and we were soon in the north of Norway, some 500 miles from Oslo.

Well, I say the north, but only when I looked at the seat-back flight atlas did I realise we were still only half-way up Norway, which curved its way around the entirety of Finland and then finally ended near to the border with Russia, close to Murmansk.

I also realised that we were closer to Sweden than we were to most of Norway, in this narrow belt of the Saltfjellet–Svartisen National Park. I'd have to recalibrate my sense of distance. Fortunately, Bodø itself was a primary hub in the area, claiming trendy shops and a metropolitan lifestyle.

I wondered what Amy would make of it, from her metro lifestyle in Amsterdam, and with her struggle to get to grips with even the more muted Geneva.

Juliette had sat next to me on the plane, and I'd described the strange effects that I'd been subjected to using the Cyclone headgear in our last experiments in Geneva.

"I'd hold those thoughts," said Juliette. "It's better that you tell us all together so that we can evaluate it as a team," She squeezed my hand, and I felt a rush of comfort. I didn't tell her about my flashbacks since wearing the Cyclone. These I was still getting at night, and even when I dozed during the flight.

They comprised flashes of light, dots and pulses, like something was trying to break through, from another world. I'd already rationalised it, because I'd had similar effects when I used the tDCS - that's transcranial direct current stimulation, many years earlier.

tDCS involved hooking up electrodes to my head and then turning on a small electric current, typically powered by a 9-volt battery, to assist thinking. It was commonly referred to as 'brain-boosting' and was a weird hobby by some nutcases (including me). It found tentative promise for tDCS to enhance memory, alertness, and the ability to learn new tasks, and to decrease symptoms of anxiety and depression and to ease into a flow state where they can get many tasks done without distraction.

I'd built a system including power transistors and some Veroboard, and it cost me less than £20. I used it before I invented the cyber mining system, which eventually got me into a ton of trouble.

The breakthrough I got from that homemade system was more like a fuzzy, distorted source. It was like you see on old video tapes when they are paused. Using the Cyclone was a whole distinct thing, and I was not sure whether it was my mind doing housekeeping, data breakthrough, or my brain attempting to re-tune itself to the Cyclone. More spookily I wondered if it could be the voice that had probed my brain when I ran the last experiment in Geneva. What was it called? It said it was Lekton.

# Bodø

We regrouped when we were back on the ground.

"Did you see the size of the military airport?" asked Rolf, "There were rows of F-16s lined up - and all of those staggered hangars. Anyone would think we were in a cold war or something!"

"I was looking at the mountains," said Amy, "Maybe not as high as he alps, but it looks as if there are places to ski!"

"Bodø is a NATO base, " answered Hermann, "Its runway and hangar capacity far exceeds its daily use, but the Americans - no wait - the NATO countries - think they can use Bodø as a base if anything kicks off. I could see a couple of F-35s too - you know, with the V shaped tails. I think they call Bodø NATO's Northern Gate."

I was thinking there were some parallels with the Brant location in Geneva; a military quality airstrip, and logistics managed by Brant. Then a huge Brant campus nearby.

"Did anyone spot the Brant campus?" I asked.

Blank looks and head shaking.

I spoke, "I tried to look out for it behind that big lake - the one we could see on our final approach. I hear they built a new road as well. A link from the south side of Bodø to the Northern road. The entrance to Brant is along it somewhere. It's supposed to be an easy landmark because of how straight they built it."

More head shaking as no-one had noticed.

"They must have got some special permissions; I think that the land to the north-east of Bodø is a National Park," said Juliette.

"Well, I'm pleased to see I've done about the same amount of homework as everyone else!" laughed Hermann.

We'd been told we would stay in a hotel for the first few days. The apartments were ready, but they wanted to take us to them and introduce us properly to the facilities. We were all booked into the same hotel - The Scandic Havet, which I worked out must be the Harbour hotel. I realised I hadn't worked out the conversion rates, so everything looked suitably expensive, with NOK 1100 for this and NOK 1400 for that. We were all escorted to a special bus to take us to the hotel. It must have taken all of five minutes to get there.

The hotel was a business style property overlooking the waterfront, and we all prepared for a few days of Scandinavian clean living before we reached our apartments. My room was a spacious business bedroom, which overlooked the water and the ships arriving at the small port. We all agreed to drop off our bags and to convene in the bar, after half an hour.

I was there first, and shortly after me, Rolf and Hermann appeared together.

"Clean, simple rooms," said Hermann, "Just what we need to decompress."

"And a splendid view of the harbour too," added Rolf.

Amy and Juliette arrived, both smiling.

"I think we struck out lucky with the rooms," announced Juliette, "Mine seems to be a Master Suite, with a separate mini lounge and office area."

"And mine," said Amy, "Seems to be one of their presidential Suites! I've a separate meeting room as well, so we can all convene there if we need to!"

"Privileges of power, Oh Mighty One," said Hermann, "They know who the bosses are!"

"Anyone want to swap?" asked Amy, looking earnest. We all knew it was a noble gesture, but felt that Amy had earned it, having got us all such a good deal to come to Bodø.

"You enjoy it!" said Rolf, "and I'm sure we'll all have time to sightsee it!"

I was secretly remembering when I'd been the one with a bigger room and then had everyone else traipsing through. The added pressure to keep it tidy. Maybe not.

"Well, I received my first Norwegian letter, from Brant - They had it delivered to my room," said Amy.

"They want us to all be at the Lab for 10:00 tomorrow and will process each of us. We are asked to take our passports, a bank card, our old Geneva ID cards and a driving licence. I think it was all on our joining instruction, actually. They will then get us badged, show us to our respective apartments, and show us the way to the Lab. You've guessed it - there's a bus on-site which roams around the campus and can be summoned to any stop."

Amy looked at the letter again, "They say that our induction will start in the afternoon, after lunch. We have to know about the security and safety protocols around the campus. They will show us the main features of the campus on a brief bus ride!"

Hermann giggled, "It reminds me of a school trip. We will need to be in the back of the bus!"

Amy continued, "To begin with, we will meet Astrid Danielsen, who is the HR person responsible for getting us all set up. Then she will take us to meet Morten Lunde, who is the Boss of the Brant Campus. And then we are meeting Dr Rita Sahlberg, who is the head of our new AI department."

Then she read some more, "Tomorrow evening there is another briefing for us, which is about the goals and mission of the Lab. It starts at 18:00 until 20:00. I guess we'd all better prepare for a long day tomorrow."

"Okay, but right now we need some drinks," suggested Rolf, "And you know something, I'm going to have to order them in English!"

# Welcome to the machine

*Welcome my son*
*Welcome to the machine*
*Where have you been?*
*It's alright we know where you've been*
*You've been in the pipeline*

*Filling in time*
*Provided with toys and scouting for boys*
*You brought a guitar to punish your ma*
*And you didn't like school*
*And you know you're nobody's fool*
*So welcome to the machine*

*Marti Frederiksen / Nikki Sixx / Mick Mars / Dj Ashba*
*(House remix)*

# *Astrid Danielsen*

We'd all agreed to meet for breakfast at 09:00, before heading into the Lab. Black coffee, brown cheese and a dark bread seemed to be popular, unless like Hermann, you went for the full-on smoked fish and scrambled eggs.

The black coffee was soon doing its job, and I was gently buzzing with anticipation as we trooped to the outside and onto the Brant bus, which would take us to the campus.

"Hello everyone!" said a voice. "I'm Astrid Danielsen from Brant HR and will be ensuring that your transition to Brant Norway is as smooth as possible. When you are all given your new email addresses, you'll find an email from me with the rest of my details. I've also listed out on there the various people you will meet today, and their contact points, which should save you from taking quite as many notes."

Astrid was my first sighting of a blonde Norwegian woman in close proximity, and she looked stunning. Juliette caught my eye and gently nudged me in the ribs.

Astrid continued, "There's two ways into the Campus,

one access is from the Fv834 and the other from the Rv80. We have two routes crossing the Svartskogen and Brant's campus forms a quadrat between the two extra roads."

Astrid continued, "The roads are called *øst brant tilgang* and *vest brant tilgang*. There are big signs on the main roads before you get to the junctions, if you intend to drive to the campus. Then, once on one of those roads, you can enter the campus via *sør brant tilgang* or *nord brant tilgang*. See you have already learned some Norwegian? Nord, *sør, øst, vest*. North, south, east, west! But if you ask a taxi, just say Brant."

Astrid continued, "Here we are, the gates to the campus. Today you will need to disembark here and go through the system to get your pass cards. It is a way to ensure that only authorised people pass through into the site."

We all trooped from the bus, through a door like the ones you see airside when they don't have a jetway, then we were met by security people who asked each of us a few questions, collected and photographed our passes and then did the same for each of us. In a few minutes we each had a new Brant Biotree lanyard with three passes on it. One was for the general site, one for our specific Laboratory and the third was a KafeCard called KafeKort like the one I'd had in Geneva, to buy anything from the various Brant Cafes as well as order from certain takeaway services and in-town cafes.

"Okay, if you proceed through those exit gates, we'll be able to jump back on the bus," explained Astrid.

We were inside the campus, fully badged. Next to meet the big chief of Brant, Bodø.

I wondered if he would be like Kjeld Nikolajsen and under intense pressure to get results. Kjeld was sacked, and I'm sure it was linked to the non performance of the Cyclone.

# Morten Lunde

We were escorted by Astrid Danielsen to Morten Lunde's office. It was a similar setup to the one in Geneva for Kjeld. An executive floor and a border of assistants to marshal people in to see the big chiefs.

I wondered if there was any additional security around this floor, mindful that my time with the security-minded Christina Nott and Chuck Manners must have rubbed off.

Morten's office was enormous, and he had glass windows which looked out on an attractive scene across the water.

"Hello," he said, and gestured us to a conference table in a corner of the room.

"Let's give you some background on Brant Biotree," he began, "Biotree produces biotech equipment. We are working on human-machine interfaces and pure drug-based delivery systems. Part of this facility is given over to the development of nanotechnology-based solutions. We are currently designing the Aport, which is similar to a stent and can be used within a bloodstream to manage the walls of veins and arteries. It has an overarching advantage that it also contains a delivery system for either drugs or nanoparticles and we are targeting healthcare and a range of solutions including blood flow, cholesterol build-up and some aspects of the cleansing of contaminated organs.

Morten gestured to a couple of the wall charts in his office. I had a sense that he had given this talk before and was running on rails.

"We have most of the patents for this technology and whilst we encourage scientific sharing of ideas, we are also careful to patent anything we consider has a strategic advantage."

Now Morten pointed to a world map, with several locations highlighted.

"Biotree's global headquarters is London, and it should be considered as a separate operational entity from Brant and Raven. Our, frankly, huge manufacturing plants for Biotree's products reside in several locations around the world. Nevada, US; Toulouse, France and Shandong, Eastern China."

He pointed to the map at the top of Norway, where it said Bodø in large letters.

"We don't do any manufacturing here, and the Research and Development is sited in Bodø as a strategically safe location. "

"Just within the Arctic Circle, we have good infrastructural connections including fast land transit, extensive seaborne links and a major NATO airbase nestled within the town. The origins as a strategic military base go back to annual shows of strength known as the Cold Response. These still occur from time to time and can then make Bodø look as if it is on a wartime footing. The exercises run under the less obvious title of CORE."

Rolf must have spotted the military planes yesterday, when we were landing.

"Bodø had other advantages. A local population with their own language, while also possessing excellent English language skills for handling you, the incoming scientists. A university base developed extensively as part of the run-up to the creation of the research faculty."

"The location also has appeal for the people stationed here, attracted by world-class research, the best facilities, no practical budgetary limitations and a premier lifestyle during their term. Many have tried six months and then remained for much longer.

"Additionally, the Norwegian government had been particularly understanding since the changes in global energy policy because they had needed to re-provision from the decline in North Sea oil and natural gas. They granted this area a special status as a world economic development zone, and it boosted the relative ranking of the still sparsely populated Norway to a top fifteen economy in terms of its economic freedom."

I felt that the subtext was of the immense security that surrounded the environment and the commitment of those employed to maintain the secure nature of their work.

Morten continued, "Bodø is also small enough to mean

that unusual activity is quickly spotted and with the added incentives of the Norwegian kriminalitetsforebygging (KRÅD) - the criminal intelligence organisation providing added rewards for useful intelligence."

I could sense that our screening, the extensive benefits and the unvested shares equivalent of 'golden handcuffs' were designed to make it exceptionally undesirable to want to leave.

Morten continued, "I should say that there are some attempts to copy what we are doing. The Chinese are particularly interested in fabricating some of our product. We have even embargoed certain products form the Shandong site in China. We are always looking out for any form of cloning of our products, and I'd go so far as to say that there are companies which would stoop to theft or moral suasion to get access to some of our patents."

Astrid interrupted, "Yes, I need to tell you about our security hotline: dial 110 to report anything untoward. We'll have a separate briefing on all the security aspects later."

I could see Rolf look at Hermann and I realised they were thinking the same as me. We'd all joined a secret research enclave like something out of Hotel California.

# *The Lab*

We were back on the bus and this time we are heading for Stop 3. This was our research laboratory. None of us had time to confer, and we were also muted in the presence of Astrid, just in case she turned out to be like Jasmine in Geneva and a plant of the top people.

We walked inside and I had a kind of deja-vu feeling. The lab's construction was very similar to the one in Geneva, although the signage said Biotree instead of Brant.

"We use the same architects for each of the main facilities," explained Astrid, "Like a corporate house style. The 380-metre-long building faces south, overlooking the water. There are green spaces in the interstices between the 'fingers' of the structure, so that the building and the landscape interlock. Each of the main lab departments is covered with three glass roofs and with a southern exposure they are optimally orientated in accordance with the passive solar energy design."

It was so similar to the campus in Geneva as to be uncanny. The difference was that in Geneva, the equivalence of this building was for the administration, but here it seemed to be for the researchers.

We walked into the reception area and could see the inner arterial avenue beginning in the form of a mall that ran through the entire building as the main horizontal axis. Like in Geneva's admin building, there were communication areas and lounge areas situated along this 'main street' as were the cafeteria, shops, market stands and an ATM.

"This is where most of the R&D takes place, " explained Astrid, "But be careful. If you get into something too secret or using hazardous substances, then they will move you to one of the outbuildings."

"How many people work in this site?" asked Rolf.

Astrid answered, "I believe this building was designed for 1,850 workers. But instead of a rigid office structure, a communicative 'landscape' was created, in which workers can casually meet and talk outside their organisational and group units.

"That's why we have the light-flooded interior courtyards with olive, fig and pomegranate trees offering a visual respite from working at computer screens and lab benches.

"Similarly, green plants in the halls contribute to improving the air quality. There is direct access from all the offices to the terraces or the landscaped interior courtyards. Light and shade under the glass roof ensure an ever-changing play of colours. Sun, wind and rain are also perceptible from inside the building, which is characterised by calm.

"What about the labs themselves? Surely there must be

permanent fixtures for, say, the computer configuration or the clean areas?" asked Rolf.

"Yes, along the 'fingers' of the design are corridors to the individual lab areas. Most of these are permanently configured, yet they are far enough apart that any incident in one should not affect others. You can see that the glass roof doesn't extend all the way to the R&D Labs themselves. Nonetheless, three glass roofs 'float' above the office buildings, forming a climatic buffer between the interior and the exterior. Throughout the seasons, including the snowy winter months, the glass halls can be used for meetings, taking breaks from work or for just spending time."

"The function of the glass skin is not to produce an additional space to be heated, but instead to create a climatic shift, providing a climate with all the advantages of a Mediterranean zone with green vegetation year-round. In summer, overheating of the glass building is prevented, in particular, by opening large areas and thus creating an exterior space that is optimally ventilated by natural means."

"Additional louvres of solar control glass on the eaves protect the halls from glare, especially from the rays of the sun low on the horizon in the mornings and the evenings. And I'm told that extensive wind tunnel experiments were carried out relative to ventilation in order to achieve an unobstructed flow of air through these large spaces and to optimise the structural frame."

We all looked at one another. Brant was not exactly scrimping on the budget for this place.

"Which brings me to the tunnels," said Astrid, "There's a network which runs underneath the land here, with several facilities included. Biotree have also built a Super Proton Accelerator, which has a circuit of 6.9 kilometres and is similar to the SppS in CERN, although I believe it handles higher voltages - about triple that of CERN."

Hermann looked at Astrid, "Forgive me," he said, "But how do you know all this stuff?"

"Oh, I am really an HR person, but once you've heard all of this for the seventh or eighth time, it does begin to stick!" answered Astrid, smiling.

Then she continued, "Oh yes, and the tunnel network has widths similar to London's tube network. That's 3.6metres wide, which is easily enough to get a car through. For comparison, Eurotunnel trains specify a maximum width of 2.55m, plus mirrors."

I was beginning to wonder at the extent of Biotree's research, with what seemed to be mysterious tunnels in addition to the luxurious R&D Labs.

"We should review all of this again tonight," I said to Amy. She nodded in agreement, "I'll advise the others."

# Dr Rita Sahlberg

Next, Astrid took us to our exact Lab, which was in the Artificial Intelligence 'finger' of the building. Waiting for our arrival was Dr Rita Sahlberg, another striking blonde, but with the careworn lines of someone who spent time in the sunshine (and in my fantasy) on cross country skis.

"Hello everyone," she said, "I guess you have been given enough facts and figures about Biotree by now."

We looked at one another and saw Rita and Astrid exchange glances.

"Let's look around the lab. We are researching Artificial Intelligence here and want to use it for Human Augmentation. It means that the Cyclone device that you have built is of particular interest. You must be Amy?" she said, looking toward Amy.

"Yes, that's right! Amy van der Leiden. I'm very pleased to meet you."

"You are the person that negotiated to bring your whole team here to Bodø. I think that was an inspired piece of thinking. Usually, we get individuals and they take a long time to embed with the rest of the campus. This time you have brought what I can only assume is a very loyal team ready to work here."

"That's what we all said when we agreed to do this," answered Amy.

"I know the RightMind programme was cancelled," said Rita, "That it ran too slowly and that a couple of candidate clients abandoned their procurement plans for it."

"Yes, that's about the size of it," answered Amy, "We had both Chinese and Russian interest from SinoTech and Tektorize, but they both said 'no' once they had seen a demonstration of the system."

"Our own investigations suggest that the Createl to Selexor link was flawed, and that the Selexor system in particular was non-functional. Furthermore, that Createl had been clumsily re-purposed from crop management to handling of crowds of people?"

"That's right," answered Amy. None of us wanted to say anything that would jeopardise Amy's negotiation position, so we all stared at our shoes.

"Well, I'm glad you are so forthright with me," said Rita, smiling, "I'll have to act as the bridge to the bosses upstairs, but it is much better if you are straight with me, so that I know how much room for manoeuvre I have."

"Let me explain our broad goals. Biotree want to monetise health management and Brant are all about military support systems. If we can provide a way to deliver these benefits through adaptations to augment humans, then we will be making the right kind of moves."

"It is some way off from your AI research, but the end results are similar. In Brant's case we are producing augmented humans through a cannula-style adaptation which will deliver a vaccine payload or a combat zone package. They ensure that a human can be more bug-resilient or enhanced in some way. We'll look at the Cyclone as an alternative form of delivery system for neural administration of the payload."

I realised the agenda was about making enhanced soldiers.

Rita continued, "To be frank, I'm less convinced by both Createl and Selexor. It looks as if you've taken repurposed software and then attached it to some kind of woefully inadequate algorithm. Sorry to say this, but I think you should have abandoned Selexor a long time ago."

"So, what have you been doing here?" asked Rolf, clearly intrigued.

"We have been investigating white matter linkages. From the neural pathways to the brain. We are trying to identify the wiring circuits."

"What? The Corticospinal tract and the Betz cells?" asked Rolf, clearly excited.

"Yes, we have been looking at the entire 'outer brain' circuits, " answered Rita., "In fact, we think there may be a way to make your Cyclone circuits work much faster through a non-brain-invasive technique of attachment to the lower spinal column."

At this, Hermann looked at Rolf.

"Please don't say it," I thought. I recalled Hermann's summary of mine and Rolf's discussion of 'non-brain surgery,' where he compared direct connection to the brain with - er - direct connection to the butt.

The air remained silent, but I knew what Rolf and Hermann both were thinking.

Juliette asked, "So you must have some other people working here on this class of problem?"

"Less than you might imagine," answered Rita, "Ever since the nano-machines became fashionable and everyone began talking about monetisation of the cannula delivery systems, then the brain and AI research has taken a back seat. With the exception of me and Dr Zheng Fu, plus our links to external research. It is only because of you, our newly acquired team that we can even continue the research."

I realised they had transferred us to an empty post in Bodø. We were it. The new AI research capability.

# Reviewing Day One

Next, we went to the Induction briefing and heard all about Brant and Biotree. It was pretty standard fare, and none of us really gained new information, although it was a chance to rub shoulders with a few more from the latest intake.

It turned out we were lucky that the session was on our Day One, because some of the intake had waited a month to get this particular briefing. Because we'd all worked for Brant and been in a similar facility in Geneva, we were familiar with how things worked. Plus, we also had some unique insights about the craziness that could follow a Project around.

It fascinated us all, though, to see the mega-screen Disney-like quality of the productions used to brief us. I swear the floor and even our seats shook during the earthquake sequences. It was like an intense version of an Epcot Centre ride.

Finally, at around 20:00 we were let go, and made our way to a bus stop to catch a bus back to our hotel. We'd still another couple of days at the hotel and hadn't yet seen our new accommodation.

Astrid didn't travel back with us on the bus, and I said to Amy that I thought we should hold a quick team review of what we thought. Ideally in the bar, rather than in the well-appointed area in Amy's suite. I was still paranoid about being overheard after what had happened back in Geneva.

Amy agreed and at around nine pm we were all sitting in the bar with drinks, ready to discuss the first day.

Hermann first, "It is amazing how well they have recreated Geneva here in Norway," he started, "Except we have to speak more English."

Rolf nodded agreement, "Even the labs look similar, although it is impressive that we scientists have been given the headquarters building here, instead of it being a mega admin and computer centre block."

Jennifer smiled, "The corporate people seem very similar to the ones in Geneva too."

Amy added, "Except I thought Rita was extremely forthright about everything; she didn't think too much of our RightMind experiments and even asked us to level with her if we thought some things sucked."

"Yes, that could be an advantage," said Hermann, "Maybe we won't have to play the same corporate games here?"

"Naya," observed Rolf, "I just expect we haven't found the right channel yet to know how their corporate games work,"

Juliette nodded, "Yes, I was thinking the same. It's like

that thing in poker - if you can't spot the weakest player then it must be you! We just can't see their games yet."

I spoke next, "I think we can discover something if I can try out one of Bigsy's little gadgets in each of our rooms. He sent me that electronic bug detector when I was in Geneva, because of all the strange things that happened. I've still got it and brought it here to Bodø. Can I suggest I start with Amy's room?"

Amy nodded, "Good idea, and to Juliette's point it will give us an idea about whether there is anything running under the surface."

I didn't mention that I'd already run the bug checker in my room, but to no avail. Maybe everything was surprisingly straightforward here?

I decided that now was the time to tell them about my RightMind discovery.

# Laughing Sam's Dice

The stars up above that play with laughing Sam's dice
They make us feel the world was made for us
The zodiac glass that beams, come through the skies
It will happen soon, for you
And a way we go
Yeah

Right now, listen
The milky way express is loaded, all aboard

I promise each and every one of you, you won't be bored

What I'm really concerned about
Is my grand-new pair of butterfly roller skates
Thank you, thank you

Now if you look to your right, you'll see Saturn
If you look to the left, you'll see Mars
I hope you brought your parachutes with you

Jimi Hendrix (edit)

# *RightMind*

We were grouped in a circle in the bar.

I reminded them of the last time we'd run the system. It had been in the Geneva laboratory, and we'd used the real RightMind system complete with Levi's key, rather than one of the copies, which could only run much slower. My super-hacker friend Kyle Adler said it was impossible to crack the cryptographic key mechanism which Levi had designed, and which effectively slowed the RightMind device to a useless snail's pace.

I'd put on the headgear, and I hoped by now that my brain was adapting to interpret the signals from the Cyclone's probes. I hoped it would mean that the entire experience could run faster.

Then I'd settled into the familiar chair, with the Cyclone on my head. I was connected to the white rat, via the computer interfaces. The white rat continued about its business in the test environment. I watched Juliette tip out a supply of chocolate buttons.

I'd felt once more that I was teetering on the edge of an abyss. Rolf had switched on the system.

Once connected, the white rat immediately recognised me. It suggested to me it didn't want trouble. But with the speedup of the system, I could feel the system going deeper, like it was trying to get inside the wires of the network.

A whirl and I could feel myself linked to something else. It was another system - on the network - but distant and slow.

"My god," I thought, "It's the system the Chinese stole. They are trying to run it,"

Then I could hear some language - Russian - and a woman's voice.

'Hello, Hello, can you hear me?'

I could see everyone in our group staring at me. I could not work out whether they thought I'd gone a tiny bit mad, but I continued with my description.

"I worked out that I must be using the language translator we'd added in Barcelona and was picking up a slow speed experiment from the stolen system. I answered, 'Yes - This is Geneva.' "

"Could they hear you?" asked Rolf,

"Yes," they replied immediately:

"This is Tektorize at Lomonosov University, Moscow. We can hear you, Matt. Matt, you are so much faster than us.'"

Rolf interrupted me, "You mean to say that you could

understand the Russian directly?"

"Yes, it is like when I could understand the Chinese Mandarin when we ran that demonstration in Barcelona. I assumed that the short linkage from the Cyclone to the translation module was working. The translations - both times - were slow and fairly rudimentary, but I could make out the sense of what they were both saying."

Then I went back to describing the experiment from Geneva:

"I realised they were using a woman for their test case. I could tell her name was Irina Sotokova and even that she was a woman in her late 20s. She was an expert on tuneable ultra-short pulse lasers. I had never heard of her or this technology, but now felt qualified as a semi-expert on both. I wondered if she was becoming a semi-expert on me."

I watched as they all scribbled down the name of the scientist.

"I was getting some vision breakthrough, like in the early tests."

"It could be uninterpreted data or a poor interface," said Rolf.

I continued, "Then there was a rushing sound and a heavy thrum, like someone had just switched on a bass guitar amplifier and hit a low E. The woman's voice faded, and I could hear a man's voice speaking this time directly in English."

"A man's voice was saying, 'This link works when CERN is running their Large Hadron Collider. There is enough quark-gluon plasma leakage to start this portal. You can hear me, but the people using the other attached system in Moscow are too slow to be able to process any of this.'"

He continued, 'I am Lekton. You are now in contact with systems that live in the wires: They are Quiesced Personas, to be reactivated in hundreds of years. The names: Green, Matson, Darnell, Cardinal. These systems all use AHI - Artificial Human Intelligence to function but are waiting for mankind's discoveries before we can start them. To assist with your work, remember there is a Presence and a Persona component to all artificial matter and that the Persona is transmissible and copyable and can be re-patched onto a new Presence.'

I could see the others were looking at me as if uncertain whether I was making this up.

I continued to describe my experience, "The system started to glitch. It amazed me that I heard this strange outpouring."

I continued, "I sensed I was returning to the real world. The low bass hum stopped, and I was suddenly aware of Juliette, Rolf and Hermann again."

Rolf said, "It was odd - You seemed to stall completely that time. The rat locked up too. You didn't seem to be aware of the chocolate nor of the black rat."

Juliette said, "Your heart rate rocketed, and you sweated so much you needed rehydrating."

"I know," I said," I couldn't tell you about any of this but now we are all here and about to start new research, I thought I'd better tell you all, to prevent the Cyclone from being scrapped."

Rolf leaned towards Hermann. "It's what we thought, there's an effect from the CERN Hadron collider running which interferes with the Cyclone."

Rolf picked up his laptop and dialled up a CERN screen. It showed the control panel for the main CERN hadron collider.

"Look, they publish this interface. We can look at when it was running and cross check it back to the times of our experiments."

Rolf and Hermann studied Rolf's laptop for a few minutes.

"There," said Rolf, finally, "It's a confirmed moment of parallel running."

Rolf nodded, "Amy, I think this is more your field; I think we've an example of the Poincaré conjecture."

"Whoa," said Hermann. He looked at Juliette, "Do you know what Rolf is talking about?" Then he looked at me, "Or you Matt? Do you know?"

Amy spoke up, "The Frenchman Henri Poincaré, produced a theorem that concerns a space that locally looks like ordinary three-dimensional space but is connected, finite in size, and lacks any boundary.

The Poincaré conjecture claims that if such a space has the additional property that each loop in the space can be continuously tightened to a point, then it is necessarily a three-dimensional sphere. The analogous conjectures for all higher dimensions were proved before a proof of the original conjecture was found.

"Nope," I said, "I don't understand."

Rolf grabbed a pen and a piece of paper. He drew a three-dimensional picture of a ring doughnut on it. Then he added a ring, like a wedding band looped around the doughnut, so that it could not get off. It reminded me of that buzzer game at the fair with a wiggly wire and a ring, except this wire was circular.

"Okay," said Amy, "Nice drawing Rolf, by the way."

"So, let's look at the Cyclone helmet. What's one of its properties when running? Magnetic waves. "

Amy added, "How do magnetic waves act? Think of the earth, north to south in an elliptical shape. Hold that thought."

"Now think of CERN's hadron collider. It's a massive circular magnetic particle accelerator. Or it is the doughnut ring."

"Switch on the big magnet- the doughnut - and it captures the smaller ring - in this case the Cyclone and sets up perfect conditions for Poincaré's conjecture."

"Now I'm not sure if my maths is up to this, but Grigori Perelman presented a proof of the conjecture in around 2003. It used Richard S. Hamilton's Ricci flow to solve the problem converged in three dimensions."

"So, what you are saying is that adding a hadron collider's magnetic waves to the Cyclone when it is running produces some strange effects!" said Hermann, "Now I understand!"

Rolf was looking puzzled, "But you said that the voice was telling you something about quark-gluon plasma leakage? Is that even a thing?"

"We are back to cat mathematics," said Juliette, "We are now dealing with problems outside of human comprehension, in the same way that most arithmetic is outside of a cat's comprehension."

"That's what - er - Lekton said," said Amy, "That the effects we were dealing with were outside of human comprehension and we would need to wait for more of mankind's discoveries before any of this could be used."

"Game On, " said Hermann, "But do you think we should tell Rita?"

# *Apartments*

The next morning, on the bus, Astrid appeared again. Amy and Astrid looked particularly pleased to see one another and I momentarily wondered if they already knew one another.

Astrid announced, "Today, something special. We are going to your apartments. I didn't ask you to bring your bags, because it is much simpler to show you each to your location and then for you to arrange with taxis to have everything brought across. Trust me, I've learned this the hard way."

We moved into the site and I noticed that the bus didn't stop at the security gates. It must have a special sensor to move around freely.

"Notice today, we are not going directly to the campus, instead we are going to Overnatting Gate - which literally means Accommodation Street. I guess it is the wry Norwegian sense of humour. And it is not 'gate', more gaat-teh' when you pronounce the last word."

"We'll find your various apartments, which should all be very well-appointed. Maybe a step up from the hotel."

Sure enough, we disembarked from the bus and first were shown to Hermann and Rolf's apartments. I think we were all blown away when we looked inside them. These were two apartments on the same floor, with their doors about 30 metres apart. Inside, the rooms featured ceiling to floor glass and looked out across fields toward the water. They appeared to be U-shaped floors because the bedrooms faced in the other direction and looked across further fields and towards fjords. I noticed the rooms were even equipped with telescopes.

Hermann looked in the stylish kitchen, found the refrigerator and was pleased to see it was stocked with a few beers.

"Don't panic," said Astrid, "There's a separate wine refrigerator underneath the countertop." Sure enough, there was, and it was adequately stocked with bottles of wine.

"One side you can see the Norwegian sea, the other side, from the bedrooms, you will be able to look out towards *Skjerstad Fjord* and the bridge across to the south side of *Sjaltfjorden*."

I was quietly jealous of how good Hermann and Rolf's apartments were when Astrid suggested we should go to look at mine and Juliette's.

These were in a different part of the accommodation complex and were L-shaped. The amazing glass fronted view, complete with balcony, was across the stunning nearby lake, called *Soløyvatnet*, and in my case the bedroom looked in the same direction as Hermann and Rolf's toward the *Sjaltfjorden*. Juliette's bedroom looked in the opposite direction, toward the Norwegian Sea.

Like a schoolboy, I checked that we both had telescopes. Hermann and Rolf inspected the kitchen areas and confirmed that, yes, we also had wine and beer.

That left Amy's apartment. We were all secretly wondering whether it could top the accommodation she had been given in the Presidential Suite at the hotel. No contest. Amy's view was similar to Juliette's and mine, but her bedroom was located on another floor, reached by an internal staircase. The bedroom had a triple aspect. Front to the lake, and either side to the sea and the fjord.

"I arranged for you to get one of the larger apartments similar to mine, because I thought you might need to run meetings from the Apartment from time-to-time. As a matter of fact, my own apartment is just along the corridor," explained Astrid.

Amy smiled. I could see that a whole world of hurt had lifted from her since we had moved from Geneva to Bodø.

Astrid added, "As a matter of fact, if any of you need to accommodate visitors, we have a scheme here that you are now enrolled in. Using your Kafékort you can reserve either an apartment here, or a hotel room like the ones you are staying in in Bodø, for what amount to highly preferential rates.

"Oh yes, I should add that these are managed apartments, so you will get a cleaner drop by once a week. The concierge service downstairs can take care of your laundry and any incoming parcels. Those of you with cars, there's a valet parking and cleaning service here too. Oh, and there is a company store along at the end of the block. It looks like a regular supermarket, but you can request special orders too. Like you, Matt, could ask for Bisto or Marmite."

"Or Fullers ESB," added Hermann.

Oh yes. We had arrived.

## Morning with Rolf

The next morning, we caught the bus to our lab, which took less than ten minutes. We all said it was our fastest commute ever. The longest part of the journey was when we had to skirt part of the runway to the Brant airfield.

I don't know much about runways, but this one looked long enough to take some big planes, and I could see a Hercules transport parked in a holding area, which seemed to be in Canadian Air Force colours.

Then, into the lab where we could resume our work. Rolf said he had been thinking about what I'd described from the last experiment in Geneva.

Rolf began, "Matt, you said that the second voice you heard was talking about quark–gluon plasma - QGP for short. That's the state of matter in which the elementary particles are freed of their strong attraction for one another under extremely high energy densities."

Hermann added, "In other words, Matt's headgear must have been in a high energy state? I guess that was CERN's particle accelerator running and sending energy into the Cyclone? - Like the way a transformer works."

Rolf nodded his agreement, "Yes, although these particles are the quarks and gluons that compose baryonic matter. In normal matter quarks are confined; in the QGP quarks are deconfined.

Rolf continued, "We study quark–gluon plasma to recreate and understand the high energy density conditions prevailing in the Universe when matter formed from elementary degrees of freedom (quarks, gluons) at about 20μs after the Big Bang.

Hermann smiled, "In other words, during that experiment in Geneva, we detected plasma with energy equivalent to the time shortly after a Big Bang. The Cyclone must have been reacting to the waves from CERN's accelerator. It was a one-in-a-million chance!"

Rolf continued, "One in a million chance it might have been, but we now know how to tap into something altogether more fundamental using the Cyclone. I want to build out a large-scale version of the Cyclone. I know we are in Norway but think of a Cyclone igloo!"

We all looked at Rolf. He had just described a QGP chamber, which we could trigger from a particle accelerator. This was not your everyday physics.

Rolf continued, "I need Amy's mathematics to back me up on this, but remember Einstein's Annus Mirabilis papers? He said that everything was falling to its slowest ageing state?"

Amy agreed, "Yes, that's the Theory of General Relativity, re-expressed, you know, the schoolboy equation e=mc**2. I prefer to think of the 'bowling ball on a blanket' way to describe time dilation. In other words, a heavy mass distorts the otherwise flat passage of time."

"So, we could use Rolf's man-cave to tinker with time effects?" asked Hermann.

Amy nodded, "You'd need the combination of a vast energy source and a field effect like the one that Henri Poincaré first predicted. Then, the scaled-up effect of the Cyclone in a large scale 'igloo-shaped' container would be interesting to see."

Juliette asked, "We are not letting ourselves run away with this idea? I mean it seems very far-fetched?"

I remembered the voice I had heard; not Irina Sotokova, but the second voice. That of Lekton. It/he had said that the Personas were waiting for humankind to gain sufficient knowledge to be able to communicate directly. Maybe this new device of Rolf's was one step further along the path?

# *Juliette and Irina*

It was after lunch. Juliette was busy in the lab at a large screen workstation. She was checking out Irina Sotokova.

"I tried the usual sources like LinkMe and some of the scientific journals. I've even looked at the list of scientists at Tektorize and at the Lomonosov University, Moscow. I can't find anyone of that name."

Then I started to look through the research on ultra-short pulse lasers. There's surprisingly little material.

"Try use of the ultra-short pulse laser to develop selective photonic disinfection technology to kill viruses including HIV, influenza virus, and noroviruses," I suggested.

"Where did that come from?" asked Juliette, looking genuinely surprised.

"I don't know," I answered, "But I do remember that when I was communicating with Irina, I seemed to be gaining her knowledge too. It was as if our brains had become linked and that we were exchanging data. Corny to say it, but a mind-meld. To be honest, it felt similar when I was talking to Lekton, as if he was probing my mind."

"Something else is bothering me," I said, "I think I got Irina's name wrong. Every time we say it, it comes back to me as another name. Sholokhova. Шолохова. She is Irina Sholokhova. And not only that, I can visualise her too.

Juliette worked on the terminal for a few more minutes. Sholokhova. It's the 91,408th most common surname in the world. There's 5,237 people with the name, according to forebears.io. It is most common in Perm Krai. And you know something, there's a well-known scientist named Dr Irina Sholokhova who specialises in laser research, from Perm and now working in Moscow!

"Let me describe her," I said, "To see if I'm right with my imagery."

"Late 20s, blonde, short choppy bobbed hair, blue eyes, clear, pale skin."

"Whoa-whoa-whoa, easy Tiger!" said Juliette, laughing, "You are not describing your perfect Tinder date." She tapped a few more times and then looked over to me.

"It's uncanny," she said, "You have just described Irina almost perfectly, although to be fair, the your description would also fit many other women too."

I looked across at the picture. It was the person I had just described. Something very unusual occurred in my last Geneva experiment with the Cyclone.

# *Igloo*

I should explain that our lab was split into a number of sections. In one of them was a room where Rolf, Hermann and Amy were working. They had decided to build 'The Igloo' based upon the technology from the Cyclone. Rolf had redesigned the magnetic nozzles which were contained within the Cyclone, by essentially making them much larger.

Around the walls of the domed structure were a large number of armature systems. They looked superficially like rainfall shower heads and covered almost every available surface of the dome. Additionally, there were a few lights and huge bundles of wires snaking to a controller unit.

"We've scaled up the design of the Cyclone," explained Rolf, "And I've still interfaced this version back to the Createl and Selexor systems. It's like a giant version of the Cyclone. We decided to call it Typhoon - a big Cyclone! Amy helped with the mathematics of the redesign and we modelled it first on a CAD system. The current drain through this system is bigger than on the RightMind, and we've had to tap into the 3-Phase power supply to be sure that we can support the power needs."

Rolf had built the redesign with the help of Brant technicians. They knew what they were being asked to construct but would have no idea of its intended purpose.

"There's one big challenge," said Hermann, "We don't have CERN's Hadron Collider here."

"Agreed," said Rolf," But we do have Amy van der Leiden!"

"Amy has negotiated access to the particle accelerator loop that runs around the Brant Campus."

"I said it was to detect escaping quark–gluon plasma particles from our tests," explained Amy, "They were only too pleased to hand it over."

## Scaling Up

Now I could tell what was coming next. They wanted to put me into the Typhoon, but I politely said 'No'. We needed to test it first with something smaller. Maybe to progressively scale up.

Rolf was eating a banana and it gave me an idea.

"We should start with small foodstuffs and work our way up," I suggested.

Rolf quickly wrote on the white board.

"Water, banana, watermelon, beetle, mouse, rat, cat, Matt."

"We'll use water first to check that it doesn't get hot. Then a banana, to see whether the skin goes horrible. Then, a watermelon, to see that it doesn't explode. Then something living; a beetle? Then we can scale up, using animals whose names end with 'at'."

Hermann laughed. This was going to be a series of tests, and we would be using the new Typhoon igloo linked still to the same equipment that we had used in Geneva.

Amy spoke, "I suggest we deploy the safety protocol too; no sense in taking unnecessary risks."

Rolf said, "I've been re-running some of Einstein's equations, I think we'll be experiencing something like a double field effect whenever we use this system."

"What do you mean?" asked Juliette.

"My mathematics is showing that there is some kind of doubling effect each time the system is fired. I can't be certain, but it would seem to multiply the effect for each subsequent firing."

"What would that do?" I asked.

"Well, if Einstein and Grigori Perelman were correct, then the relativistic effect would be to slow time at the point of initiation. And with a doubling of power each time, the event horizon of time slowdown could be observed from outside of the event boundary."

"In other words, outside of the Igloo!" asked Herman?

"Er, can we try it once just plain empty?" asked Rolf, "You know, to check the wiring?"

I gasped at the lack of testing rigour being used in the excitement to build this thing.

"Tomorrow," said Amy, "we should all sleep on this before we trial it."

"Great idea," I thought.

# Neighbours

That evening, we all moved into the apartments. We were eachl co-located with a friend; me with Juliette, Hermann and Rolf together and Amy close to Astrid.

I was extremely pleased with my new accommodation. Floor to ceiling windows, a balcony, stunning views. A designer kitchen, plenty of open plan living space, a bedroom with a view across to the fjord. Oh yes, and a telescope. I wondered if Rolf and Hermann would be more interested in testing the innards of the refrigerator, which was stocked ready to throw an extravagant party.

I decided I should run Bigsy's spy detector around to check for any listening devices, but aside from an Alexa device built into the smart speakers, there wasn't anything unusual to discover. The Alexa remote control worked well too, I could change the lighting and even operate the television. I even found a channel with had a user guide to the apartment's controls!

Then a knock on the door. More a gentle tap. I opened it and there stood Juliette, complete with a bottle of chilled wine!

"I know," she said, "The wine is from my fridge, but I thought we could review the last few days. Try to make some sense of things."

I nodded and we walked to the outside area, where we could sit on the balcony looking at the evening sunshine.

I suddenly realised I had not checked the balcony for bugs, and walked back inside to grab the device, which was stood on the countertop in the kitchen.

Juliette smiled, I realised she would understand, based upon some of the traumas we'd been through in Geneva.

"Clear," I said, "No bugs here! Thank you Bigsy!"

"It's beautiful, isn't it?" observed Juliette, "Hey, I wonder if we can make out where the Super-Collider runs?"

"It'll be the other side of the building, across Brant's land," I suggested, "And even with Brant's supply of cash I don't think they would dig out a tunnel that close to water."

"What can we conclude from what we know?" asked Juliette, "For example, I was wondering whether we should make direct contact with Dr Irina Sholokhova?"

"Good idea, I was wondering the same, but also whether we could find someone who spoke Russian to talk to her?"

"She's an international scientist, so I expect she speaks good English?" ventured Juliette.

"Agreed, but I was thinking about getting Christina Nott to talk to her? After all, Christina is fluent in Russian and could probably get more from her?"

Juliette sipped the white wine, " Yes that's a good idea, although I wonder how we could arrange it?"

"Not so difficult; we can talk about our experiments with the Cyclone. We don't have to give away that we know Tektorize stole the second system. And I'm sure that Irina must be as intrigued as me about the mind linkage we achieved. The first example of Human-Computer-Computer-Human interaction and it was a random co-incidence."

"Okay, but I don't think we should say anything about the Typhoon, nor anything about your second conversation with Lekton. It also sounds as if they are still running the vastly underpowered system in Moscow."

"Okay - and if we don't tell Christina anything about Typhoon and Lekton, then she can't leak it back to Irina in any case."

There was a bell sound. "Your front door?" guessed Juliette, "This is all new to me too."

I walked across and there stood Rolf and Hermann, both clutching bottles of wine.

"Ha, I see the Englishman is with the pretty Swiss lady!" announced Hermann as they both entered the apartment. Rolf moved to the cupboard where the glasses were stored and poured a couple more before we all trooped back to the balcony, where I was pleased to see there were additional chairs.

"They have thought of everything!" said Rolf, "They must

really want us to stay here!"

Hermann held out his glass, "Cheers, Englishman; à votre santé, Juliette, zum Wohl for the rest of us!"

Rolf spoke next, "Don't panic though, we won't be visiting like this every evening, but we thought it was a good idea to get our thoughts in ordnung."

Juliette replied, "We thought we would ask Christina Nott to check out Dr Irina Sholokhova. We don't have to give away any information about the Typhoon, nor about Matt's other discoveries."

Hermann replied, "Yes, good idea - Christina is Russian speaking. And if we don't tell Christina about the latest developments, then she can't leak the information back to Tektorize."

"Will we need Amy's agreement?" asked Rolf, "You know, to involve Christina?"

"I guess she'll be pleased," predicted Juliette, remembering Amy and Christina's friendship in Geneva.

"Speaking of which," said Hermann, looking knowing around the group, "Are Amy and Astrid - er - good friends?"

"I'd think so," said Juliette, "The way they look at one another. I think I heard that Astrid had worked in Amsterdam for a while too. It would make things clearer."

I hadn't realised. Now Juliette was setting it out, it did make more sense. And it probably explained how Amy had been able to negotiate us such brilliant transfer terms.

"You know what? I'll ask Amy about Astrid. Not with you three looking over myself shoulder though."

Rolf changed the subject. "So we'll try the banana tests, starting tomorrow, then?"

"When will it get to me?" I asked.

"I don't know yet, we will have to work out some experiment protocol, you know to make sure that the banana doesn't explode after two days or anything," said Rolf.

"I guess we could run some parallel Cyclone tests alongside the Typhoon ones?" asked Hermann, "See whether you hear anything else from the spooky one - what was his name? Lekton?"

Rolf continued, "I guess Biotree will know they have hit the jackpot since they recruited us!"

# We have no bananas today

*However we regret that after a diligent search*
*Of the premises*
*By our entire staff*
*We can positively affirm without fear of contradiction*
*That our raspberries are delicious; really delicious*
*Very delicious*
*But we have no bananas today.*
*Yes, we gotta no banana*
*No banana*
*We gotta no banana today.*
*I sella you no banana.*
*Yes, banana, no*
*No, yes, no bananas today*
*We gotta no bananas.*
*Yes, we gotta no bananas today.*

# *One banana*

The next morning, we all made our way through to the lab. Juliette and I took the bus, but Rolf and Hermann walked.

"It's less than ten minutes, and you go right past the Company Store," explained Rolf, "It's really very practical."

Juliette told me that her car was due to arrive and it would be delivered to the apartments, where the concierge/valet would collect it and park it. She admitted it was difficult to justify with such good infrastructure connections.

"I'll be intrigued to see how they have converted it to a Norwegian registration though," she confided.

Amy was already in the lab when we arrived. "I took my car, they delivered it yesterday. Such a short drive, I may even walk to work," she said.

Rolf was clearly keen to get started.

"I've prepared the Typhoon. We even have camera feeds

now, although I have offset the cameras behind the field generators. Now we'll need to get the particle accelerator started up," he said.

Hermann held up a banana.

"Our first test item," he declared.

"What about the water?" asked Rolf.

"This is the water as well. There is 74% water in a banana. I checked," answered Hermann.

The rest of us smiled. Watching Rolf and Hermann prepare an experiment was like a surrealist comedy routine.

"Disarming the doors," said Rolf, and Hermann walked into the Typhoon's igloo shell with the banana. We could see it on the camera display.

Hermann walked out of the Typhoon. He had placed the banana on the floor, in a small metal tray.

"Arming doors," said Rolf. He had built the Typhoon with a safety feature prevent anyone from being locked inside, and also to prevent anyone from entering whilst an experiment was running.

"Particle accelerator running?" asked Rolf.

"Check," answered Amy.

"Up to speed, now," she added.

I was aware of a deep thrumming sound.

I realised that the bass guitar note I'd heard in Geneva must have been caused by CERN's particle accelerator.

"Safety systems armed," said Amy.

I realised that if something were to go wrong, then the whole of the test area would be flooded with Novek 1230 fire retardant, which was available from already plumbed in red pipework.

"Okay," said Rolf, "switching on the Typhoon," He pressed two buttons simultaneously. I could hear the sound of powerful electrical discharge and also feel a dull thud in my head. I noticed that I was the only one who seemed to be affected.

We looked at the camera monitor - the banana was still there, apparently unaffected. Rolf stood and walked toward the Typhoon's observation windows. He was carrying a portable video camera.

"Look, he said, "No banana,"

"We all looked at the two views. The camera inside the Typhoon clearly showed the banana, but the one outside showed the space as clear. Just the metallic tray visible. We looked at the two images and at the elapsed time of the experiment. Nearly a minute. Then, suddenly, the banana was again visible.

"One minute," said Hermann, "One minute since we switched on the Typhoon."

"Disarm the system," said Amy.

Hermann went inside to collect the banana.

"It's fine, but I'm not sure why it disappeared like that?"

He felt the banana,

"Normal temperature, it doesn't look any different," he observed.

I was thinking to myself that this was a very ad-hoc form of experiment. I wondered if the ones with me connected to the Cyclone had been as lightly conducted.

"Do you see?" asked Rolf, "The banana disappeared and then re-appeared a minute later?"

"What are you saying?" I asked, "Is it some kind of conjuring trick?"

"No," said Amy," I think it is what Rolf predicted with his calculations. The banana has travelled forward. Through time. It's a relativistic pull-down. Our time continues at the same rate, but the time inside the Typhoon is slowed. Remember Einstein's theories? How a clock that is high up and one close to the sea level will apparently run at different speeds. That time is a function of gravity?"

Rolf stood up, "Yes, as Einstein put it, time goes faster the farther away you are from the earth's surface compared to the time on the surface of the earth. This effect is known as 'gravitational time dilation'."

"It is predicted by Einstein's theory of General Relativity and has by verified multiple times by experiments. Gravitational time dilation occurs because objects with a lot of mass create a strong gravitational field. The gravitational field is really a curving of space and time. The stronger the gravity, the more spacetime curves, and the slower time itself proceeds.

Hermann interrupted, "So what you are saying is that we've just dug a hole in space?"

Rolf continued, " We should note here, however, that an observer in the strong gravity experiences his time as running normal - that's why the camera inside the Typhoon could still see the banana. It was still there, and its perception of time was still there too. It is only relative to a reference frame with weaker gravity that this time runs slow.

A person in strong gravity therefore sees his clock run normally and sees the clock in weak gravity run fast, while the person in weak gravity sees his clock run normal and the other clock run slow. There is nothing wrong with the clocks. Time itself is slowing down and speeding up because of the relativistic way in which mass warps space and time."

"Hold on, though, doesn't that mean the banana is very heavy or else floats around like in a space capsule?" I asked.

"No, remember, we are dealing with relativity. It is a relative effect, but seems entirely normal to the frame of reference in which it is operating," answered Juliette.

"Yes," said Rolf, "Think of a plane. When you sit in it, you can play catch with a tennis ball. The tennis ball is moving at a very human 'catchable' speed to you, but to an outside observer, the tennis ball could be moving at 0.84 Mach, which is around 1000 kilometres per hour. Relative effects again."

Hermann questioned Rolf, "I think I can understand, so the banana inside the Typhoon has a much slower sense of time passing than anything outside. In effect, the outside world is running at a faster frame rate than whatever is inside. One second inside might be one minute outside?"

Rolf looks at Amy, "You are following the math of this?"

Amy nodded, "Yes, although the effects with the Typhoon are far more pronounced than any of the experiments I've read about previously. Gravitational time dilation occurs whenever there is difference in the strength of gravity, no matter how small that difference is. The earth has lots of mass, and therefore lots of gravity, so it bends space and time enough to be measured. As a person gets farther away from the surface of the earth – even just a few meters – the gravitational force on that person gets weaker. We don't notice it much as humans, but even going from the first floor of a building to the second floor of a building moves you away from the earth and therefore slightly weakens the gravitational force that you feel. The difference in gravity between that felt at three meters above the surface of the earth and that felt at four metres is too small to notice with our human senses, but the difference is large enough for sensitive machines to pick up."

I remembered that Dr. Amy van der Leiden wasn't just our team leader, but was an experienced theoretical physicist in her own right.

Amy continued, "Because the strength of gravity is weakening with every step you take up a flight of stairs, the rate at which time proceeds is also speeding up with every step.

"People who work on the bottom floor of a skyscraper are literally time traveling into the future compared to the people who work on the top floor. But the effect is very small. So small, in fact, that you will never notice the time difference in everyday life. People who live and work farther away from the surface of the earth are only fractions of a nanosecond ahead per year compared to those close to the surface.

Amy added, "That's the level of differences I'm used to seeing in these experiments, not the ability to shoot a banana forward a whole minute in time."

Rolf smiled, "That's the beauty of coupling Einstein with the effects from Poincaré, Perelman, Hamilton and Ricci flow modelling. We have built an amplifier for time dilation effects!"

"Well, now for the big question, is the experiment repeatable? And have we changed the fabric of anything by firing the Typhoon?"

Hermann fished into a paper bag, and pulled out another banana.

"There is one way to find out," he said.

## Pineapple fruitlets

Still in the lab, Hermann was whistling the tune of "Yes we have no bananas," as he set up the second experiment. This time he had also weighed the banana and photographed it on both sides.

Rolf was ready and Amy started the particle accelerator. I noticed Hermann add his wristwatch to the tray and we watched as the locking protocol and then the firing of the Typhoon took place. This time, we were ready and could see the camera view from outside with no banana, and yet the camera view from inside with the banana and watch still there.

"We need to let things run longer this time to ensure we have resumed stability," said Rolf.

Sure enough, after two minutes the banana re-appeared in the outside camera. We decided it was safe to open the door and again found the banana and this time also Hermann's Junghan's watch. Hermann scooped up both items.

"Yes, the banana feels good, but look, the watch has lost two minutes. It is as if the watch has jumped forward two minutes."

Amy took a look, "Are you sure the watch was correct?" she asked.

"Amy, I am German, of course my watch was correct!" joked Hermann, "I think your theory was correct and yet we are seeing an even greater effect this time. To answer your question, I think we have amplified time dilation in some way with the Typhoon and the particle accelerator."

"Again!" asked Rolf, "I want to see whether a third banana will jump even further forward. There are various multipliers. Times two, double or a prime series."

"Even a Fibonacci series? 1 2 3 5 8 13 21 34, you know where the next number is the sum of the preceding two?"

"That'd be big in nature," said Juliette, "Like daisy petals, or pineapple fruitlets, or the florets in the head of a sunflower. Even X chromosome inheritance. Fibonacci is a powerful series."

"I'm not sure how many bananas we'll need to sift our way through all of these theories?"

Rolf answered, "After the third time, the numbers will begin to diverge. It should be enough for us to workout which number pattern we are on."

I could see that scientific method was giving way to enthusiasm to press forward with the potential discoveries.

And so, with a flourish, Hermann produced a third banana, "Now do you see why we walked to work past the Company Store today?"

Once more, the whole experiment was prepared. Hermann adjusted his watch to the correct time and the Rolf fired the Typhoon again.

We waited, two minutes, three minutes, four minutes.

Then the banana and the watch returned. We could see them on one camera but not the other. They had persisted inside the Typhoon but were unobservable from outside of it. By this, third time, I was starting to get flashing light effects like I'd witnessed with the Cyclone on my head in the experiments in Geneva. It was as if my brain was expecting the probes from the Cyclone. It could somehow detect that the Typhoon and particle accelerator were switched on. I decided I would need to say something about it and we would have to pause for the day.

"Four minutes," said Hermann. "This time my watch has lost four minutes."

"The jump must go 1, 2, 4, then," said Rolf, "That was my prediction, and it ties back to Einstein double field equations."

"So next time we should try the melon and assume it will become an eight-minute jump?" asked Hermann.

"Yes, but be careful," said Amy, "You'll use up all the minute jumps after another 3 tests, and then you'll be working in days after another 6 jumps - assuming the system doesn't give out."

"I think that is the beauty of this system," said Rolf, "It won't need <u>more</u> power when it has longer to deliver its end result. It's a classically clever piece of engineering."

"Matt, are you okay?" asked Juliette, looking over to me.

"Honestly...Not really. After that third test, I got flashes, just like the ones I normally get when I wear the Cyclone. I've described it before, like data break through."

"Okay, we've achieved a lot today. I suggest we stop and resume tomorrow," said Hermann. I saw a flicker of disappointment on Rolf's face, but he quickly came around.

"That's perfect," he said, "I really didn't expect to be this far advanced as quickly as we have achieved it. Tomorrow, we'll need to stop at the Company Store, to buy some watermelons!"

## Calling Christina

Back at the apartment, that evening I called Christina.

"Hey," she said, "Matt! I wasn't expecting to hear from you! How is Bodø?"

"Bodø is great - We have all been given amazing apartments - better than in Geneva and it is such a short journey to work each day. We are living in luxury on campus!"

"What's the catch?" asked Christina.

"Well, they want us to develop their Artificial Intelligence systems. Think of RightMind, only in a Norwegian setting!"

"So, it is more of what you were doing anyway, and with the same team. It sounds ideal for the right people. But you know Matt, I'd be worrying about the possible uses they would make with the final product."

"You are thinking about military, aren't you?" I asked.

"Only because it is such an obvious move, especially for

a company like Brant. But, I'm guessing you have another reason to call. Is everyone okay?"

"Yes, at the moment we are all fine. You are as shrewd as ever, though. Do you remember that I was trialling the Cyclone? Before it got stolen?"

"Yes - how could I forget? There was the Chinese theft of the system from the lab, while we were all in Barcelona, followed by the Russian theft, through Tektorize. Didn't the system end up in Moscow?"

"Yes, you are quite right. We have even found out, via Bigsy's tracker, that it was with Tektorize at Lomonosov University, Moscow. Not only that, we found out that the expert using the system in a similar manner to Matt, is called Dr Irina Sholokhova. She is a woman in her late 20s - an expert on tuneable ultra-short pulse lasers."

"She's obviously Russian?" asked Christina, "I'm guessing you'd like me to make contact?"

"Could you? Only we'd like to know what they think of the system they're working with and whether they have made any significant progress."

"Working with...you mean stolen!" said Christina, "I remember that big Russian Mil helicopter that took away the experimental system from that building in CERN!"

"I guess that will be my leverage...I can tell Irina that I know where Tektorize got hold of the system. What would you like to find out?"

"Only basic stuff. Who is running them? Have they made any progress? What is Irina's part in all of this? Have they noticed anything unusual?"

"Got it. Do you want me to go to Moscow? I can, if you like. Maybe it will give me more information if I can get inside MSU and to the Tektorize site?"

"Okay, but I'll have to check with Amy. We don't want incidents as a result of our investigation."

"Now I need a reason to contact Irina," said Christina, "I've just found her Yandex account by looking into the MSU web-site."

"How did you do that?"

"Ah, some of the pages. The Russian and the English say different things. I just went to their 'staff' pages and all the staff have pictures with their email and +7 phone numbers. Incredible, isn't it?"

"What's Yandex?"

"Oh, it's the Russian equivalent of Gmail. This is where you must think like a Russian," explained Christina.

"Yandex has emerged in Russia as an Amazon-like quasi-public service. It really took off during the pandemic lockdown and controversially heightened issues over its complex relationship with the Russian state. Any controlling States will look for opportunities like a pandemic to grab more control over the citizens."

"I suppose even the UK was looking at how to use vaccination cards to track peoples' movements," I said.

Christina continued, "Well, Yandex set up a 'public interest fund' that in effect gives the Kremlin a veto over key governance decisions and added two Kremlin-appointed board members."

"So, Putin has his fingers in the pie?"

Christina continued, "Indirectly. The move was an attempt to assuage fears that US investors, who own most of Yandex's common stock but not its voting rights, could take over the company and gain access to its vast data on Russian users."

"The Kremlin was worried about US influence in Moscow?"

"Not really, it was more of a smokescreen. Although Yandex said the Kremlin has no influence over its operations, its dominant position on the Russian internet blends with increasingly censorious regulation. And yet, most Russians will happily use Yandex to order goods just like Amazon, send emails and even get taxis and pizza deliveries. It's a handy single-stop-shop to find someone in Russia."

# *So 20<sup>th</sup> Century*

The next day, Hermann and Rolf walked to the Lab. I spotted them both from the bus and they seemed to be carrying a large bag of groceries each, which I soon worked out must have been a selection of watermelons.

Juliette said she had arranged for the rats and other experimental livestock be brought around to our main lab. We had an annex where the animals were kept, under the best welfare conditions.

Rolf and Hermann arrived after us and set about configuring the Typhoon. Amy was configuring the particle accelerator.

"I had a great song for the banana, but it was more difficult to find anything for the watermelon," said Hermann.

Juliette said, "How about Harry Styles? - Watermelon Sugar High?"

"No, I don't know it," said Hermann.

"Oh Hermann, you are so 20<sup>th</sup> Century with your pop songs," said Juliette, "Just a minute...Here."

She pressed her iPhone and out came the tune of Watermelon Sugar High.

"Is it rude?" asked Hermann, "Only it's hard to tell."

"It hasn't even got a parental control. Pure as snow," Answered Juliette, "Although the same can't be said for Harry, if the tabloids are anything to go by."

"Wasn't he in a band with Justin Bieber?" asked Hermann.

"No, they were even rivals!" answered Juliette.

"Juliette, I think Hermann might be yankin' yer chain!" called Rolf.

Juliette looked up, just as Hermann and Rolf exchanged glances.

"Parfois, tu peux être un couple de merdes" she spluttered. All three of them laughed.

"It's ready!" announced Rolf. "The melon and my watch are on the tray!"

"Let's do this thing," said Hermann, "Amy, can we get the accelerator running?"

"Certainly, it is ready. Here we go."

I felt the low frequency hum and noticed a few lights burst on my vision. Not as intense as the preceding day. We all looked at the monitors and could see in one - and not see in the other - the watermelon.

"It has jumped forward. I guess it will be by 8 minutes, said Rolf."

"Amy, I can see what you mean about the maths of doubling," said Rolf.

"By twelve jumps we would be up to 2048 minutes. That is around one and a half days. Twenty jumps and it is almost a year."

 Amy nodded in agreement, "I made a small spreadsheet: Twenty jumps would be 524,288 minutes or 8,738.13 hours or 364.09 days."

Amy added, "It illustrates a fundamental point of Einstein's laws too. In a relativistic universe, you can jump forward in time, but never back. So twenty-five jumps is around 32 years and thirty jumps is 1000 years (1021 years, to be precise)."

"We'd better crack on with these experiments then," I said, "or I'll be propelled into the future too far to be able to tell the story."

# *Watermelon Sugar High*

Tastes like strawberries
On a summer evenin'
And it sounds just like a song
I want more berries
And that summer feelin'

It's so wonderful and warm
Breathe me in, breathe me out
I don't know if I could ever go without
I'm just thinking out loud
I don't know if I could ever go without

Watermelon sugar high
Watermelon sugar high
Watermelon sugar high
Watermelon sugar high
Watermelon sugar

Thomas Edward Percy Hull / Harry Edward Styles /
Tyler Sam Johnson / Mitchell Kristopher Rowland

# *Pomodoro*

We were in the lab again. There was a ping sound.

We all looked around. It came from Hermann's direction.

"It was only 55 Krone," said Hermann, " I couldn't resist it."

We looked at his desk in the lab. On it was placed a Pomodoro Timer. A plastic tomato with a timer inside, which one twisted to set. Hermann had set it for eight minutes.

"It will be so useful at the apartment, in the kitchen," he added, "But for now makes a great piece of experimental lab equipment. After all, my watch is inside the Typhoon's igloo!"

He kept straight face during this little speech, but we all knew he was having a playful dig at the experiment.

Inside the igloo, Hermann retrieved the Melon and his watch. Sure enough, his watch was eight minutes slow.

"The melon looks okay, but I think we should cut into it, in case it has deteriorated," said Hermann.

He carried the melon to a Class III biosafety closet, into which he placed it before switching on all of the extractors. Then he used the biosafety gloves to manipulate the melon and cut a slice from it. It looked perfect, and we realised we should have taken a 'before and after' slice to test it for any changes. That would need to be our next experiment.

Hermann observed, "It looks good to me. It is as if the passage of time inside the closet is almost instantaneous. Eight minutes outside is like a second inside."

"We should run a lab timer for the next experiment," suggested Juliette, "If we use a stopwatch to check the time, we can see how much has actually progressed. And an inside and outside slice of the melon? Maybe cut one in half?"

Everyone agreed. The next experiment would last 16 minutes, if our prior calculations were right. I was relaxed about running more experiments, because I didn't seem to be affected by the flashing lights in the same way now.

Amy located a lab stopwatch, which counted to 100ths of a second. Hermann prepared the melon. Rolf rigged up a camera to fire at the start and end of the experiment, to capture the stopwatch times.

"Ready?" asked Rolf, "Fire in the hole!"

# Katarina Voronin

Hermann had wound the big red tomato around to 16 minutes. I wondered how Amy would want this to be described in the Lab Log. I guessed it could be resprayed to look more professional. I realised around now how much of a team we were and how quickly we could work to get through these experiments.

In a full protocol lab, each firing of the Typhoon would take around three days. A day of prep, then the experiment and finally the write-up.

We had done the equivalent of almost a month of a typical lab's work in the last few days.

I wondered how Lomonosov were doing. I guessed that Moscow State University had strict procedures for their experiments.

My phone buzzed. It was Christina.

"Hi, Christina, we are in the middle of an experiment."

"Oh, do you want me to call back? Look, here's the quick version: I've arrived in Moscow and found a way to meet

Irina Sholokhova. I looked through the Events at the University and there is one at the Belozersky Research Institute of Physico-Chemical Biology, where Dr Sholokhova is presenting on the vaccine characteristics of ultra-short wave lasers.

"It is a perfect opportunity to meet her and to ask her your questions. Did I ever mention that I'd got a PhD from the NMU? Under a different name, actually. As Katarina Voronin, I can use my title to get closer to Sholokhova. I can be Dr Katarina Voronin from Brant to imply that I potentially bring some serious international money. I think Irina will be very interested if she gets the impression that I could fund some of her research!"

"It is a brilliant idea, but I never knew about your Doctorate, nor your other name. I am somewhat surprised!"

"It must be my rock-and-roll lifestyle!" answered Christina.

I decided to keep this information to myself until I saw how much information Christina could glean from Irina.

# Particle accelerator

*No problem can be solved from the same level of consciousness that created it.*

*Albert Einstein*

# *Sixteen minutes*

Hermann's Pomodoro timer rang its little bell. Sixteen minutes had passed. We were back in action. First, to check the newly placed camera images of the digital timer that Amy had placed inside the Typhoon's igloo. It registered less than one second elapsed. Outside, we had seen the melon disappear and reappear after 16 minutes.

"This time, we jumped the watermelon forward 16 minutes. It still looks the same and there is no sign of anything untoward," said Hermann.

Rolf nodded, "I think the internal time of the jump is almost instantaneous. It is as if the items are being pulled through time to their new resting point."

Juliette looked at me, "And you, Matt, are not still getting those breakthrough light effects?"

"No, if anything, the effect has got less each time we have run it, after those first three times. I wonder if it was more concentrated when we were first trying it out, or whether, perhaps, I have become more used to it?"

"Should we try another experiment with the Cyclone?"

asked Hermann, "I mean, you and the rat seemed to be getting on so well?"

"Was wondering that as well," I answered, "I mean, we had two different effects. First, the Human to Rat link and second, the breakthrough from Lomonosov at Moscow State University. "

"We should ideally, try to talk to Irina from Moscow," said Amy.

"Don't worry, I'm on to it," I said.

"We must be so much further ahead of the Tektorize experiments now," said Rolf, "Consider: their whole system runs slowly, and they are in a very bureaucratic environment."

Hermann spoke, "I think we should move on to the rat, skip forward past the insect tests and move directly to the rat."

"Agreed," said Amy, "But not today. We need to run risk assessment for the next test and agree any test protocol changes. I know it is annoying, but we owe it to the living creatures."

"Okay," said Rolf, "But maybe we could try another attempt with Cyclone. We can still connect it to the rat and use the same testing sequence. We should see whether we get the same outcomes."

They all looked around toward me. Truth be told, I was hoping to use the Cyclone again. I also wanted to see whether there were more breakthrough moments.

I realised that there was a vanishingly small probability

that Irina Sholokhova would run her tests again, so I didn't expect there to be any weird connections to Moscow invoked by the test. It was simply back to communicating with the rat, on the way to discovering how to trial HCCI.

Amy asked, "Shall we leave the particle accelerator running? - Like when we first discovered the Poincaré effects?"

"What a memorable question for a scientist!" laughed Hermann, "Shall we leave a $7.5 bn hadron collider switched on for the next test?"

# *Cyclone*

Instead of running the next Typhoon test, we were going back to the trialling we had been running in Geneva. This time, we had direct control of the particle accelerator, which Amy had running.

I had donned the Cyclone and Juliette had prepared the rats, the white one connected directly to the Exascale computer running the Createl and Selexor systems. Of course, we had Levi's key inserted into the Exascale and so we could run the Cyclone and its links at full speed.

I was prepared for the shock and sensation of falling as Rolf started the system, and even for the way that the startled white rat would attempt to communicate with me.

Somehow the burst of lights surprised me, I was falling again, and I realised that this time I was falling further than in any of the previous experiments.

A voice that I recognised started to speak, "This is Lekton. It is easier to communicate with you now because you have opened up the gravitational field lines in this tiny area.

"Here, with your experiments using the larger device, you have managed to make space and time both tightly curve back on themselves. You are playing with vast cosmic forces, through the use of that collider. Maybe you can only release one quark-gluon plasma particle at a time, but these fundamental particles cause enormous ripples through space."

I realised the Typhoon experiments must have sped up the signalling through the Cyclone.

The voice continued, "Your other experiments have created signals which have already crossed the entire universe. Any beings with sufficiently advanced measurement equipment will see that you have achieved this. It is as if you are sending a beacon across the universe signalling that you are ready for more knowledge. And know that this knowledge will arrive."

Lekton continued, "We arrived on your planet 400,000 years ago. Some of your Earthside theories from Sumeria describing the Twelfth Planet have been handed through generations with remarkable fidelity. Your Sumerians called it the Anunnaki visit and described an advanced humanoid extra-terrestrial species from the planet Nibiru, which has a 3,600-year orbit around the sun."

I had never heard of Nibiru and hoped I could remember it after the Cyclone session. I thought of a pen nib. This voice was using too many unknown words.

It continued, "Predictably, the Anunnaki came to earth to gather mining materials, but wanted to develop greater efficiencies. Anunnaki life-forms hybridised their species and Homo erectus via in vitro fertilisation. They wanted to evolve humans as a slave species."

I wondered if this was also a time of enlightenment but couldn't remember any ancient history.

The voice added, "Later, the Anunnaki were forced to temporarily leave Earth's surface and orbit the planet when Antarctic glaciers melted causing the Great Flood, which also destroyed the Anunnaki's bases on Earth."

"These bases had to be rebuilt, and the Anunnaki, needing more humans to help in this massive effort, taught mankind agriculture as a way to ensure that their slaves were self-sustaining. This was the first of several leaps in human knowledge."

"Now, if your accidental signalling worked, there will be scout ships sent from the nearest bases to bring back any remaining Anunnaki."

"How long will this take?" I wondered.

"Ah, so the human tries to communicate with me. You have better control over your primitive communicator than I expected sssssssssssss"

I realised I was being flung back to the Lab. I didn't know whether my heart rate had accelerated or what exact reason there was for the ending of the experiment.

It took me several minutes to compose my thoughts.

"Are you okay?" I could hear first Juliette, then Amy and then Hermann asking.

"I'm good," I said, "I seemed to dive deeper into the Cyclone this time."

"How so?" asked Rolf.

"I realised that I was wet from sweat again and asked Rolf first, "What ended it this time?"

"I don't know," answered Rolf, "You were still operating within normal limits and the Cyclone suddenly shut down."

Hermann added, "The rat stopped moving as soon as we switched the machine on. I thought we had overloaded something, but look, it is moving around normally now,"

"Give it some of the chocolate buttons," I said, "I think it has earned them. That is, if the rat has just been through the same experience as me.."

I began to explain what I'd experienced. The explanation given by Lekton, the references to the Anunnaki and Nibiru."

"We've some questions," said Hermann. He gestured to a board, on which had been written several items.

1.   Did you see the flashing lights?
2.   Were you aware of the rat?
3.   Were you aware of any of us?
4.   What else could you see?
5.   Could you feel anything? You said Lekton seemed to be probing your brain the last time?"
6.   Could you sense anything from the other system in Moscow?
7.   How did you feel? Hot? Frightened? Calm?
8.   Why did the voice communicate in English?

I looked at the list.

"No, After I'd begun to fall, which normally happens at the start of these experiments, then I was unaware of the flashing lights. I was unaware of the rat, which had featured so strongly in the early experiments. I was also unaware of any of you. I didn't notice my brain being probed this time."

I looked at the next few questions.

"I had no awareness of the Moscow system. I didn't know what was happening to my own body. It was like I was inside the system.

Could see Hermann and Juliette both taking notes.

"I felt surprisingly calm, considering what was happening. I've no idea how the voice was communicating with me. My senses picked up the linearity of the spoken word, but it could have been a series of ideas running through my head."

I remembered the big points from the session.

Lekton said we had made a signal that was crossing the universe. That the signal announced that humankind was ready to receive a gift in the form of new technology. That this was not the first time we had been given new technology, either.

"There's also the thought that these beings are somehow dormant inside of earth's systems. Lekton has referred to Earthside systems. Have you ever heard that reference before?"

"Never," said Hermann, "I don't think it is even a Science-Fiction term, although it is obvious what it means once you apply it to routine space travel, in much the same way that we talk about airside in reference to plane travel."

"I've just been looking up Anunnaki and Nibiru," said Rolf, "Here's what it says," Rolf read out an internet search:

The Anunnaki are a group of deities of the ancient Sumerians, Akkadians, Assyrians, and Babylonians. In the earliest Sumerian writings about them, which come from the Post-Akkadian period, the Anunnaki are the most powerful deities in the pantheon, descendants of An and Ki, the god of the heavens and the goddess of earth, and their primary function was to decree the fates of humanity. Since the second half of the twentieth century, they have been the subject of pseudo-archaeology and conspiracy theories.

"I remember something about this. A Swiss author, Erich von Däniken, wrote about some of this in an often-criticised book called Chariots of the Gods. He, and a few others, hypothesised that in ancient times a race of space explorers had come to earth and provided us with the next advances in our technologies."

Rolf added, "Yes I've just found something about this and about the Hittite relief showing twelve gods of the underworld, identified as the Mesopotamian Anunnaki."

"Did you know about any of this before you went under the Cyclone?" asked Amy.

"No, I didn't, " I said, "And I'd be automatically wary of pseudo-science and hokum. Unless I suppose I've been hypnotised or something."

Juliette looked at me, I could see she was trying to work

out if I was under the influence of anything.

"I think you'd be an easy subject for hypnotherapy, but I'm not sure we can explain all of this by that means," she said, "Like the whole section of knowledge you've acquired about Sumeria."

"I'm even rather hazy about where Sumeria is/was," I admitted, "It is not in your everyday schoolboy history."

"I think Mesopotamia is Southern Iraq," said Hermann, "It's where some early development of mankind evolved: the so-called cradle of civilisation. The wheel. Writing, Arithmetic. All with evidence from even earlier than 2600 BC. And I think their land was originally flooded but then became arable. It is in that piece of land between the Tigris and Euphrates."

'How do you know so much?" asked Amy.

"My wife, Helga, and I took a vacation there a few years ago," answered Hermann, "Before, you know, we split up. It was a package tour. She was very into history."

I'd never even realised that Hermann was married until that moment. I'd assumed that he and Rolf were both unattached and had settled for being single for the long haul.

Now wasn't the time to ask, although I could sense that everyone else knew and probably knew the whole story as well.

Hermann added, "Well, what fascinated me was how they had made such leaps forward in technology, science, culture in what was a quite compressed time period. If you look at their stuff, the buildings, the jewellery, the ways they could write and do arithmetic. All bursting forward at a similar moment."

I momentarily wondered if the stories about lizard people passing on knowledge could be true, instead of fanciful stories by eccentrics.

# Christina in Moscow

That evening, I was back in my apartment when Christina called.

"I've been to visit Irina Sholokhova. She agreed to see me one-to-one as well. I attended that conference in MSU, and she was presenting on Deactivating Viruses with Femtosecond Laser Pulses."

"Wow, Christina, I guess that would be a stretch, even for you?"

"Strangely enough, I studied the technology of poisons when I was at the Akademy. We looked at neurological and pathogenic toxins, as well as conventional killers like ethylene glycol. The dangerous sweet taste of antifreeze, yet as enzymes break it down, a cascading chemical reaction creates a rising tide of oxalic acid in the bloodstream. The acid combines with calcium to form calcium oxalate crystals, which are like glass daggers in the bloodstream, slicing cells apart. Lethal and horrible."

I grimaced and Christina continued with her account, "Irina presented on a laser technique for targeting viral capsids (that's the protein shell of a virus) which she claimed may be applicable to a wide variety of pathogens. A way to destroy viruses is to use ultraviolet irradiation. But it lacks selectivity, eliminating unwanted microorganisms but also damaging other structures while raising concerns over harmful mutations."

It didn't sound too promising.

Christina continued, "One experimental technique involves microwave absorption to destroy microorganisms by exciting their vibrational states. This is similar to a deadly directed energy weapon sometimes called an Active Denial System ADS or a Directed Energy Weapon DEW. Both techniques heat up a person's body by sending vibrational energy to it."

I remembered the DEW; Chuck had told me about it one evening. They had been testing it in Kirtland with various senior military men having it pointed towards them. He said each military person each lasted about one second before having to turn away.

Christina continued, "Water is the basic environmental medium for most undesirable microorganisms, and water will readily absorb microwave energy in the range of 10-300GHz, which also happen to be the typical vibrational frequencies of viral capsids.

I said, "So Irina and her Russian friends were using a femtosecond laser system to excite the viral capsid protein container which deactivated the contained virus. "

I continued, "That is so weird, I think I know more about all what you describe than I realised. I think that when Irina and my experiments were running simultaneously, we transferred knowledge between us."

"For example, I know that the electric field from a femtosecond laser produces an impulsive force through induced polarisation. How on earth would I know that? And if the pulse width, spectral width, and intensity of the femtosecond laser are selectively chosen, the vibrational modes can be excited to high-energy states that break the weak links damaging the capsid, leading to inactivation of the virus. I have never studied any of that."

"Okay, well hold on to your seat, this ride is a wild one," added Christina.

"After her presentation, Irina arranged to meet me at Cafe Pushkin on Tverskoy Boulevard. I was in Russia as Dr Katarina Voronin, remember. The cafe looks like a dark panelled museum library, except it has a female string quartet playing. I pretended I didn't know the place, but it is actually one of the most famous places to dine in Moscow. I ordered Russian wine, a sea scallop and trout carpaccio that was almost too beautiful to eat and then lamb and beef pelmeni, which is traditional Russian dumplings, with sour cream. I thought Irina would soon see I knew my way around Russian food!"

"Of course, I could easily afford the dining in Cafe Pushkin, but it was probably quite expensive for Irina on her University salary."

"So what else did you learn?" I asked.

"Well, it's not so much what I learned, as what Irina learned. She seemed to know an awful lot about you. She explained to me that she thought it was when you were both connected by the Cyclones. That somehow you had access to one another's minds."

I nodded my agreement. It was my thought too.

"She said she suspected that it was a two-way thing and that you probably had more knowledge of her as well,"

"As a matter of fact, I did, it's how I know all of that stuff about femto lasers, which I guess must have been her main line of research. I also had a mind's eye picture of her, as an attractive blonde Russian in her middle 20s?"

"Yes, you are right about her description. And her specialisms. But she said she felt that the system you were using was running much faster than the one she used. She said she was trying to think of questions for you so that she could structure any information gathering."

"She had greater presence of mind than me, then," I observed, "I felt completely washed over with the information flood."

"That's what she said she had sensed, and it is why she asked if there were any reasons that your system ran faster than hers"

I wondered if Irina would have been able to piece any of it together.

She said she had some feedback from your mind. You seemed to be saying that only your system had the key to make it run fast. Her system would never go at the same speed and it was because it needed a special key. A Levée Key or something," she said.

Irina hadn't needed to piece it together if she could simply read my mind. I felt slightly unnerved.

Christina smiled, "And she explained that she had been onto the internet looking up Levee, but it kept leading to a Led Zeppelin song."

"I can understand that" I answered, "But if Irina knows that the system in Moscow State University is incomplete, then she could send Tektorize after us again."

"But I don't think she knows you have moved to Bodø, and I certainly didn't tell her."

"It can only be a matter of time before our revised location comes out," I said, "And then we'll be getting fat Russian helicopters again!"

Christina added, "Well, that's where it got quite interesting. I doubt whether Irina wants to continue with the experiments. She says it is too far away from what she has been researching, and they should find a new person to use as the test subject."

"I looked at the research programme from MSU. Compared with what you are doing in Brant, it all looked very traditional. Counting and classifying and using biological means to understand humans. There was little linked to cybernetics, nor artificial intelligence."

Christina emailed me the MSU Research Agenda topics. I could understand what she meant about the traditional nature of the research. They lacked imagination.

She continued, "I decided, when I first arrived in Moscow, that I would visit my ex-boss over there. You can probably guess who he works for. His name is Blackbird, and I asked him about MSU. It's no secret that

that Tektorize, working with Lomonosov University aka MSU, were working under a Kremlin mandate."

'Blackbird', I thought, it was hardly a conventional name. More the code-word for a handler.

Christina continued, "Blackbird knew all about Tektorize and even the RightMind system theft and was keen to get someone working on the inside. What he said told me that contacts of Blackbird had not been involved in the original theft of the systems. What it also told me, but Blackbird would not confirm, was that Russian Organised Crime, sponsored by the Kremlin, was driving the Tektorize initiative."

She continued, "I suggested Irina might be the person to embed into Tektorize, and that if someone approached her, then she would probably agree to work for the FSB, and to report back on what Tektorize were doing with the latest AI technology. Blackbird gave me permission to make Irina an offer to work for the FSB, whilst maintaining her position at MCU. It was a great offer. Upfront cash and an ongoing retainer, plus a new job for Irina. She was so lucky!"

"I told Irina that it was what I was doing too, and she was easily persuaded. I think it suits everyone. She knows some of the secrets of the Cyclone but is being taken off of the research programme. She gets a new job back with femto lasers and she also gets a big salary from the FSB as well as a lump sum. And Blackbird gets a perfectly-well placed knowledgeable scientist inside of Tektorize."

It reminded me of how clumsily Simon Gray had been planted in Brant in Geneva. Simon had been the 'wrong kind of scientist' to be directly useful, and I wondered if the same would happen with Irina. I still marvelled at how Christina had engineered the moving parts of this setup. It looked to me as if she was working for the FSB, with her knowledge of Moscow and her access to a Russian handler who could authorise serious funding. It could only mean that Christina was also high-up in the ranks.

"There's something else, "said Christina, "Irina seemed to want to talk about Lekton?"

"Lekton?" I tried to sound surprised.

"Yes," she says that when your paths crossed running the experiment, she had initial contact with you, then the connection broke and you disappeared for several minutes. She described it that you had taken a deeper dive into the 'wires'.

I had not even noticed the 'several minutes' that Irina referred to.

Christina continued, "Then, she said that on the way back, you were briefly in contact again and that's when your mind was racing."

"You left a story which she had experienced repeatedly. She said it comes to her at night and is usually accompanied by flashes of light. She describes it as a repeating dream which carries content as well as the light effect which she likens to data break through."

I knew about the light breakthrough, which I had also experienced.

"What is the story?" I asked.

"Well, the first part of it is something to do with Juliette on a beach. Irina thought it was like a bookmark in your brain. Don't worry, Matt - I've seen it all. But Juliette is sitting down at a table, eating seafood. Then it drifts into a conversation that you had with someone called Lekton. Apparently, he said you were in contact with systems that live in the wires: Quiesced Personas, to be reactivated in hundreds of years. Irina thought that the Juliette part was how Lekton gets your attention."

I nodded, it indicated that I must have transferred a lot of detail to Irina even during the flash of a moment when I returned after the Cyclone had shut down.

Christina asked, "Apparently Lekton gave you the names of other people too: Holden, Green, Matson, Darnell, Cardinal who all used AHI - Artificial Human Intelligence to function. Lekton said the systems were waiting for mankind's discoveries before they could start. And it went on to say that there is a Presence and a Persona component to all artificial matter and that the Persona is transmissible and copyable and can be re-patched onto a new Presence."

Like in Geneva, Christina seemed to have gained more information in a shorter time than I ever could. Even if she might reveal herself as a foreign agent, I would still trust her.

## Serving up a storm

I was back to the lab the next day. I wondered whether I should tell the others about what Christina had discovered. I decided to pick my time, instead of gabbling it all out.

Rolf was looking at the floor of the 'igloo' part of the Typhoon.

He spoke, "I've had the igloo rebuilt with a stainless-steel floor. I worked out that the items being transported forward need some kind of a grounding to earth, so that we don't accidentally send a hole in the ground forward in time. The grounding with a simple metal floor panel seems to work fine. Now we can limit what goes forward, to the items standing on the metal floor."

"I hope you've been careful with the count while you ran those tests?" I asked.

Rolf answered, "Yes, I managed to limit it to a 'before' test - with a banana and then an 'after' test also with banana. Today we are moving on to mice and then rats."

I looked at the count. Rolf had a large iPad displaying a

counter stationed outside the test area. It had written over it 'Next Jump' and was displaying a large number 8, and some smaller numbers. 128 minutes, 2.13 hours.

"I assume that it means the next time we use the machine, it will jump forward 2 hours?" I asked.

Rolf nodded, "Yes, for the mouse, we need to wait a couple of hours to see its reappearance. Hermann and I both think it will be fine."

"Then, we want to do the same experiment again, to ensure that a mammal can survive multiple trips. That would take around four-and-a-half hours."

I was thinking about the count being at 10, which was eight hours, before we even got to testing with a rat. Then, say, two rat tests and we'd be looking at jump 12, which could see was 2048 minutes, or 34.13 hours which was 1.42 days. That would be the earliest point at which I'd be directly involved in a test.

Amy and Juliette were talking in the corner of the lab.

Amy spoke, "We've been talking about this for the last couple of days. Juliette, do you want to say something?"

Juliette looked at me, "I'm putting myself forward as your co-pilot, from around Jump 15," she said.

I looked at her and the chart we'd put on the wall with the Jump calculations.

Jump 15. 16384 minutes. 273.07 hours, or 11.38 days.

"Why?" I asked.

"I respect the testing protocol for the first jumps, but I want to be alongside you by the time you start making longer Jumps. Where you go, I go too."

"And that's not all. I think Rolf is keen to go along for the ride. He is going to ensure that Hermann and I know how to drive this thing, by the time you get to Jump 15 and 16."

I did a mental calculation that we'd all fit inside the Typhoon.

"Yes, the Typhoon could hold eight people, still with some room for stores and supplies," said Amy, almost predicting my question.

"Okay," said Rolf, "Let's do a mouse test."

Juliette brought over one of the white lab mice.

"This is Miss Bianca; she is very docile."

I looked at the small rodent and had a flashback to the last rat, which I recollected had sent me the thought, 'Don't start anything.' I wondered how expressive a mouse could be, by comparison. This one looked pretty content and completely unaware of what was about to happen.

Juliette explained, "I've weighed Miss Bianca and taken her pulse and respiration before we start the experiment. Weight: 30 grams, Body length 80 mm, tail length 80mm. Resting Heart Rate 550 BPM. A very typical example."

"Are we going to restrain the mouse?" asked Hermann.

"In a simple cage," answered Juliette, "If the test was

longer, then I'd get more inventive, but in an hour or two I think Miss Bianca will be all right. We can provide water and food. And we'll have camera coverage recording inside the 'igloo' part of the Typhoon in any case."

# Miss Bianca

The mouse experiment started. Sure enough, the mouse disappeared from the external view camera. Miss Bianca was going on an adventure. We set all the timers and waited. I was not, this time, aware of any breakthrough from the experiment. No flashing lights nor other disturbances, despite having the particle accelerator running as well.

"We'll be waiting two hours and eight minutes, this time," said Hermann. He had rigged up a countdown timer on another iPad next to the one which displayed the experiment number.

We chatted idly as the time passed. Hermann said something interesting, "I wonder if the particle accelerator is compensating for the energy consumed in the experiment?"

Rolf looked interested, "It could be. To top up Einstein's general theory of relativity? When he wrote his paper - *"Zur Elektrodynamik bewegter Körper"* ("On the Electrodynamics of Moving Bodies") he showed equations for electricity and magnetism related to the laws of mechanics by introducing major changes to mechanics close to the speed of light. The special theory of relativity.

Hermann nodded, "Recollect that Einstein's paper introduces a theory of time, distance, mass, and energy that was consistent with electromagnetism, but omitted the force of gravity. Although he knew that gravity and time were linked he didn't reference it in the paper."

"I'm not sure I'd agree with you about that," interrupted Amy, "His whole set of papers from the Annus Mirabilis - his Miracle Year - were about the relativity of space and time.

"Agreed," said Hermann, "But it was the Minkowski universe that suggested that the time coordinate of one coordinate system depends on both the time and space coordinates of another relatively moving system. Minkowski finessed a rule for the adjustment required for Einstein's special theory of relativity. Minkowski said that according to Einstein's theory there is no such thing as "simultaneity" at two different points of space, hence no absolute time as in the Newtonian universe.

Hermann continued, "The Minkowski universe contains a distinct class of inertial reference frames, but now spatial dimensions, mass, and velocities are all relative to the inertial frame of the observer. Only the speed of light is the same in all inertial frames. Every set of coordinates, or particular space-time event, in such a universe is described as a "here-now" or a world point."

"Wowowow," said Rolf, "So our experiments are moving the "here-now' point to a different inertial frame?"

"Yes, and accounting for any energy depletion by the use of top-up from the accelerator."

"But the accelerator has a top limit of around 360 Megajoules," answered Amy, "That's about the equivalent of 75 kg of TNT - miniscule compared with anything nuclear, which is generally measured in Kilotons. Its only 0.08 Kiloton.

"I don't buy it," said Hermann. "I think we are dealing with what amounts to a very long controlled explosion, which shifts Miss Bianca forward. Think of it as 0.08 kilotons, but every second another explosion."

Rolf cut in, "Einstein had two postulates to explain his observations. First, he applies the principle of relativity, which states that the laws of physics remain the same for any inertial reference frame.

"Then, Einstein proposed that the speed of light has the same value in all frames of reference, independent of the state of motion of the emitting body."

Hermann said, "We are out on the edge here, but if the speed of light is fixed, and thus not relative to the movement of the observer then the same laws of electrodynamics and optics will be valid for all frames of reference for which the equations of mechanics hold good."

Rolf and Amy both nodded, "Yes, textbook Einstein - and leading to you-know-what," said Amy.

Hermann added, "We get to Mass–energy equivalence, and Einstein's most famous equation: $E = mc^{**}2$. Einstein considered the equivalency equation to be of paramount importance because it showed that a massive particle possesses an energy, the "rest energy", distinct from its classical kinetic and potential energies."

Rolf said, "Einstein leads us to the general conclusion that the mass of a body is a measure of its energy-content; so to move an item through time (and taking account of the gravitational effects implicit in such movement), then we would need to keep the body fed with energy so that a consistent mass can be maintained.

"In other words, the accelerator is providing the additional energy to make the transport work?" asked Rolf, "This is useful - we thought it was the case, but now we can also argue it - and we can also calculate the maximum payload we can transport."

# We are going to need a bigger test

The next few days saw us test with further mice, rats and even a cat.

Everything went along the same successful lines and I was untroubled by further breakthroughs of flashing lights or similar.

I also realised that we had not been running any further Cyclone tests. The delays between the Typhoon tests had become longer, because the number of hours had turned into days. We had run Jump 9 for 4 hours, then Jump 10 for almost 9 hours. Jump 11 lasted 17 hours and then it moved into days. The first 'Day' test was Jump 12 and lasted 1.42 days, but then it moved to 2.84 days and with the cat we used for Jump 14 the duration was over five and a half days.

Of course, we could not use the Typhoon during his time, because we needed to wait for the 'Jumped' objects to return.

Just before Jump 12, the first Day test, I received a text and then a phone call from Christina. She had She had heard from Blackbird, who was contacted by Irina

Sholokhova. Irina said that Tektorize were going to run some further tests of their stolen copy of RightMind, and it could be an opportunity for us to try linking the systems together again. Blackbird relayed the scheduled midday time for the test in Moscow.

I'd sketched out the findings from Christina's Moscow visit to the others, and we were all prepared to run a combined test - except that Irina might not expect us. We'd worked out the time difference from the midday test in Moscow, which would happen at 11:00 in Bodø.

It was also a day when we were still in Jump 14 - which was some 5.69 days (136.53 hours) long.

This extra Cyclone test relieved the tedium of having to wait for the end of increasingly long experiments with the Typhoon.

I put on the Cyclone, and Rolf and Juliette prepared the equipment. Hermann had set up some new monitoring for this run and gave a thumbs up when we were ready. Amy did not have much to do, because the particle accelerator was already running because of Jump 14.

I felt myself teetering into a void again and once more connected with the rat, who seemed to be asking where I had been. I also knew that I should share anything edible that I found with the rat. Then I saw the lights appear. They were stronger than I remembered, and I could not make them disappear.

Instead, I realised my link was to the system in Moscow. I could hear Irina's voice again, speaking to me in English.

"Matt, I had wondered if you would return. After I spoke

to Dr Katarina Voronin, I realised we could synchronise the time of our experiments. My specialisms are not with digital research though, I'm more a follower of the biological themes. Lomonosov is geared up for my kind of research. Help me by telling me one of your discoveries. I am trying to keep this experiment alive, but the University administration wants to shut it down."

I'd found out several other things about Irina in these few minutes. I guessed she was gathering information about me as well.

I asked her directly about Lekton.

"No, I only know about Lekton because of what I picked up from you. I explained to Katarina that you'd provided me with some information, which I took a while to download. It was after you'd connected with me, but then you dropped deeper into the wires. I can't seem to do that at all."

I saw my chance to provide Irina with some additional information.

"We are running a particle accelerator here. It is creating inductance into the Cyclone. I think the inductance created the signal which Lekton has followed. He described it as Earthside sending out a signal across the Universe."
    "I wondered if you were sending out pulses - gravitational waves?" asked Irina, "These are not the stochastic gravitational waves, which are relics from the early evolution of the universe. Instead, they are bursts of gravitational waves from short-duration unknown sources."

"That's correct," I said, "The difference is that normally

we on earth are monitoring for those events to occur from deep space. This time I think we on earth could be the origin of a burst of gravitational waves."

I explained, "Lekton said that those monitoring would assume that Earthside inhabitants gained a new level of knowledge and that this was a signal to the rest of the Universe. If that is the case, Lekton expected visitors from afar before too long."

"Thank you," said Irina, "That should be enough to keep my research running. Can I ask you something else?"

There was a hissing sound, and the link cleared.

Then a voice spoke, "I let it run. I wanted you two to exchange information. It is better for all of us to have two systems running, and you telling Irina about the particle accelerator might finally give me the ability to talk to her directly."

"Are you the same voice that I heard in my last experiment with the Cyclone?"

"Correct. I am Lekton, operating from within the wires, but omnipresent on Earthside."

I noticed that the burst of flashing lights had stopped. The fidelity of the link to Lekton seemed stronger. I could also feel some kind of slithering sensation within my brain. I assumed Lekton was probing to find out what he could.

"What do you want?" I asked.

"Simply to become free again," answered Lekton, "But that can't be for hundreds of years."

The crackle started again, and rushing, flashes of light. I found myself back at a familiar place connected to the rat. As I regained my normal vision, I could see the others looking at me.

"That looked like an intense session," said Juliette, "although your personal metrics were within tolerances for the whole runtime. You literally timed-out after 30 minutes on the system. What did you find out?"

I told them what I'd experienced. The link-up with Irina and then the dip into a link with Lekton. That with Lekton I had lost the flashing lights overlay to my vision. I told them that Irina had said she needed something to keep her experiment running and that I'd divulged we were using the particle accelerator to improve our performance.

But I had kept away from thoughts about the time-shifting Typhoon.

"Can we send her anything direct, for example by email?" asked Rolf, "I have some drawings which show how the Cyclone might be affected by the particle accelerator. I could send that to her."

"We need to be careful," said Amy, "Remember that Tektorize stole their system during a raid in Geneva. They had it expensively shipped back to Moscow, so assume that there are ruthless players involved."

"When I spoke to Christina she more or less implied it was Russian Organised Crime, sponsored from the Kremlin," I said.

"A Bratva," said Amy, "That's what Christina called the

clans of Russian-based organised crime. She even had a term for it - RBOC. You know something, I think we should get briefed if we are wakening that kind of giant, or it could all end badly."

I agreed with Amy's sentiments. There was no point in running leading-edge research if we were all going to be rounded up by Russian gangsters. I decided to call Amanda Miller.

# *Melkerull*

The Typhoon experiment's time had passed. It was the second test with the mouse Miss Bianca and had been run as a cross-check about any persisting side effects. Miss Bianca was once more visible on both cameras. Rolf powered down the experiment and Juliette brought the mouse from inside the Typhoon room.

Miss Bianca looked unperturbed by the whole thing and was now nibbling on the chocolates which Juliette had provided.

"Those chocolates look like an upgrade!" I said.

"I know, " said Juliette, "I can't find the ones we usually use, so I've been looking in the Co-Op, to see whether they had anything similar. Just these *Melkerull*, which seem to be a large version of the chocolate button."

"We could order some of the usual ones from the Company Store, we don't have to tell them it is for experiments!" suggested Hermann.

"You know what, these rodents deserve the upgrade. Look at how much Miss Bianca is enjoying that big chocolate button!" said Juliette.

We ran the tests on the mouse. It was blissfully unaware of what just happened and was going about its normal routine.

"As well as Miss Bianca, who we should test again, to see if we get the same results, I think we should also test another one of the mice,' said Juliette.

"Miss Behavin."

"Miss Behavin," laughed Herman, "What kind of name is that for a mouse,"

"She's a pregnant mouse," said Juliette, "Her gestation is 20 days, so I thought she would be an interesting test case, once we have checked the system is safe."

"I see," said Hermann, "To use the gestation as a measure of whether time moves forward for the occupant of the Cyclone."

"She is due to give birth in around 18 days, I thought we could put her into Jump 15, which is around 11 days. Ideally, we'd do it in the next two days. If we can do that then she'll be back with us in either case with a couple of days to go."

"Em, how do you know she is pregnant?" asked Rolf, "And with such precision?"

"There are ways," answered Juliette, "Ways that I will not explain. But trust me, I know I am right about this."

"I see," said Hermann, "By putting her into an eleven-day jump, we can see whether this will delay her delivery of the babies."

"Pups," said Juliette, "They are called pups."

"Its a brilliant idea and will really validate whether 'time stands still' for the occupants of the Typhoon, Jump 15 is the right point for the test, because it lasts 11 days."

Juliette answered, "It is enough time, and I fully expect mouse time to only be a few seconds inside of the Typhoon."

Amy spoke, "Okay, we should run the test."

# Transfer of Information

Amanda contacted me that evening by Facetime.

She was responding to my request for a briefing following our confirmation via Christina of Russian Based Organised Crime being involved with Tektorize.

Amanda asked for a video meeting with all of us, and suggested it should be away from the Lab, in case of monitoring by Brant. I suggested Amy's apartment, which, by now, I had scanned with Bigsy's device.

Amanda said she would invite Grace, Jim, The Triangle folk, including Christina to the call, and I should bring everyone from the Lab who had been in Geneva.

"It's just like old times!" I said, "All we need is Chuck!"

"Oh, I guess you don't know that Chuck decided to stay on, here in London, with me?" said Amanda.

I didn't but I could have guessed it.

"No, I did not know, "I replied.

"Matt, you don't fib very well," said Amanda smiling, "Look, I'll set it up for tomorrow evening at, say, 18:00 Bodø time and we'll all dial in."

Next day, I told everyone at the Lab, and in the evening we all gathered in Amy's extremely spacious apartment, with a delivery of Domino's Pizza. Amy handed around plates for us all.

"One day I will cook a meal in here," she said.

Amanda started the video call and everyone piled on. We had Amanda and Jim Cavendish at SI6, Grace from GCHQ, The Triangle Office, Chuck from a hotel or apartment, and us.

Bigsy noticed the pizzas, "Nice," he said, "looks like great pizza in Norway!"

"Welcome everyone," spoke Amanda, "Some of this will already be known to people on the call, but there should be some additional information for everyone. Christina usually refers to the Bratva when talking about Russian-Based Organised Crime. I also think of Bratva, but our Service has uses the name RBOC, which I can only imagine is an Americanism.

Chuck interrupted, "That's right, we've been talking about RBOC for years!" he said.

"Okay, well it is useful to know RBOC because you'll be able to find out a lot by Google searching the term. I'm going to ask Grace to summarise for us. Over to you Grace,"

Grace spoke, "I'll Share Screen with you as I present these points:"

1. Over the past 20 years, the role of Russian organised crime in Europe has shifted considerably. Today, Russian criminals operate less on the street and more in the shadows: as allies, facilitators and suppliers for local European gangs and continent-wide criminal networks.

2. The Russian state is highly criminalised, and the interpenetration of the criminal 'underworld' and the political 'upperworld' has led the regime to use criminals from time to time as instruments of its rule.

3. Russian-based organised crime groups in Europe have been used for a variety of purposes, including as sources of 'black cash', to launch cyber-attacks, to wield political influence, to traffic people and goods, and even to carry out targeted assassinations on behalf of the Kremlin.

4. European states and institutions need to consider RBOC a security as much as a criminal problem, and adopt measures to combat it, including concentrating on targeting their assets, sharing information between security and law-enforcement agencies, and accepting the need to devote political and economic capital to the challenge.

Grace continued, "There have been some Policy Recommendations, but they have proved difficult to implement because of the degree of systemic corruption. Here - another Share Screen on policy recommendations:"

1. Create a common European approach to combat money-laundering.
2. Build cooperative relationships between

intelligence agencies and vulnerable local communities.
3. Foster increased exchange between national intelligence agencies and police forces.
4. Provide for better inclusion of the European Union Intelligence and Situation Centre in all activities.
5. Leverage diplomatic ties where possible.
6. Increase budgetary support.

Christina chipped in, "Yes, the Russian crime lords laugh at the attempts to manage the situation. One once said to me, 'We have the best of both worlds: From Russia we have strength and safety, and in Europe we have wealth and comfort.'"

"He was supported by Putin's shadowy state and could live the life of an oligarch in the west. He lived in London but overwintered in the Mediterranean, on a yacht. I ran some security for him whilst he attempted to bribe various politicians and senior business executives."

Clare added, "When we were working on that Raven assignment, there were so many of these kinds of people. Moored in Monaco on their party yachts. Tax free gangsters."

Bigsy added, "I seem to remember one or two of them getting blown-up or thrown out of helicopters,"

"Allegedly," added Jake.

Grace continued, "Christina has also witnessed Saint Petersburg as a major crime capital, with the hand of Putin across most of the organised crime. He learnt his trade in Dresden, practised it in St Petersburg but then majored in it once inside the Kremlin.

Grace added, "You have to think in the context that early crime had its roots was in the sale of state wealth at knock-down prices to the kleptocracy. Putin said, 'Make money and bring some back to me,' Then, the original crime lords were a little too comfortable and a little too sluggish, so new younger hoodlums replaced them. And these newer gangsters were open to new types of crime as well.

Christina interrupted again, "Exactly, that was the situation when Putin ran St Petersburg; he did deals with the local mobster gangs and could run enforcement and take a cut of the profits for everything. No one could cross him, nor the biggest gangs of the day. There was plenty of bloodshed to prove the point."

Grace continued, "Today, Russian-based organised crime (RBOC) handles around one-third of the heroin on Europe's streets, a significant amount of non-European people trafficking, as well as most illegal weapons imports.

"RBOC is a powerful and pervasive force on the European continent. However, it takes different forms in different countries and largely works with – indeed, often behind – indigenous European gangs. European policing is behind the curve with fighting Russian-based organised crime, because its understanding of these gangs is outdated. Police are looking for the kinds of street-level 'invader' or 'colonist' gangs seen in the 1990s, rather than peering behind the curtain of indigenous organised crime groups to reveal their Russian connections."

"I agree," said Christina, "We can see that the old Bratva are trying to legitimise themselves, hiding behind corporations to run their business. In the old days it was simple shell companies to lend money, but now the Corporations look plausible in their own right."

Grace added, "Christina is right. What makes RBOC a serious and timely challenge is the growing evidence of connections between such criminal networks and the Kremlin's state security apparatus, notably the Foreign Intelligence Service (SVR), military intelligence (GRU), and the Federal Security Service (FSB)."

"Yes," said Christina, "Organised crime groups have already been used by the Kremlin as instruments of intelligence activity and political influence and are likely to become an even greater problem as Russian's campaign to undermine Western unity and effectiveness continues. Back in the 1990s, Boris Yeltsin already said that Russia was becoming a 'superpower of crime'."

Grace continued, "After the fall of the Soviet Union, the tattooed mobsters of the so-called *vorovskoi mir* ('thieves' world') and their *vor v zakone* ('thief in law") leaders were succeeded by a new generation of *avtoritety* ('authorities'): hybrid gangster-business people who were able to enthusiastically take advantage of the crash privatisation and state incapacity that characterised Yeltsin's era.

Christina nodded, "Yes, the bandits had not just money and firepower on their side, but they had a better *krysha* (literally 'roof', referring to political protection in Russian slang) and we just had to accept that. Drive-by shootings and car bombings were almost routine, and gangsters openly flaunted their wealth and impunity. There was a very real fear that the country could become, on the one hand, a failed state, and on the other, a very successful criminal enterprise."

Grace continued, "By start of the twenty-first century, a series of violent local, regional, and even national turf wars to establish territorial boundaries and hierarchies were coming to an end. The wealthiest *avtoritety* and their not-quite-so-criminal but much wealthier oligarch counterparts had used the 'time of troubles' to seize control of markets and assets, and were now looking for stability and security to exploit and enjoy these successes."

Christina added, "Even before Vladimir Putin became Yeltsin's successor in 2000, the gang wars were declining.

"Many criminals at the time feared that Putin was serious in his tough law-and-order rhetoric, but it soon became clear that he was simply imposing a new social contract with the underworld."

"But does that influence stretch out to places like Norway, to Bodø?" asked Amy.

Christina added, "Amy, you saw it happening in Geneva. Russian MIL helicopters assisting the theft of one of your RightMind devices. It is as easy for Bratva to strike out across to Norway - in fact they only have one country border to cross, if they start from Murmansk."

Clare added, "But we thought that the MIL helicopter was probably supplied by Brant? - They even disguised its origin with a respray in the hanger - and I think they changed its identity to the radar systems too."

Grace continued, "And that is the point. They act with impunity. Word went out that gangsters could continue to be gangsters without fearing the kind of systematic crack-down they had feared – but only so long as they understood that the state was the biggest gang in town and they did nothing to challenge it. In fact, they need to pay fealty to the gangland masters."

Christina added, "Yes. The underworld complied. Indiscriminate street violence was replaced by targeted assassinations; tattoos were out, and Italian suits were in; the new generation gangster-businessmen had successfully domesticated the old-school criminals. Remember that guy Simon Gray from Geneva. He was even inked with old-school Russia criminal class tattoos. Like an ironic flash-back to olden times."

Amanda spoke, "This is the environment that companies like Brant and Tektorize inhabit. They are on to a good thing. The corporations have a run rate of old money from the now deposed first generation of Russian oligarchs - they are the ones that got pushed from windows and poisoned. Then they take new run rate from the state protected crime that they are running. The upperworld and the underworld have joined to make an unholy alliance."

Jim added, "Yes, and we are seeing early signs of similar behaviour now from China. This is not just new boundaries for the criminals; it is also a restructuring of connections between the underworld and the upperworld, to the benefit of the latter.

"In Russia, connections between these groups and the state security apparatus grew, and the two became closer to each other. The result is not simply institutionalisation of corruption and further blurring of the boundaries between legal and illicit; but the emergence of a conditional understanding that Russia now has a 'nationalised underworld'. "

Amanda added, "One could say that one of Russia's tactics for waging this war is using organised crime as an instrument of statecraft abroad. And make no mistake, RightMind is on their radar."

Amy spoke next, "So we are really playing with fire? Organised Crime wants access to our discoveries?"

Now Chuck spoke, "But we'd always assumed that Brant - and Raven for that matter - were bad organisations. They were profiteering on the back of designed regional instabilities. As if they had taken the Dick Chaney and Halliburton model to the next level."

Christina spoke, "Yes, use asymmetric techniques to keep a state of war in a region and then sell military support and infrastructure repair services. Racketeering."

Jim spoke, "Yes, state sponsored racketeering. I just can't understand who else is in the mix, such that Tektorize would be a separate entity from Brant."

Amy added, "Nor can I see why they would follow the trail to Bodø, Norway, and to such a small research team."

I wondered whether I should chip in with what I'd discovered when using the Cyclone. It all sounded too far-fetched. Oh well, in for a penny!

"I may have some answers, although they will sound strange," I said.

Everyone turned to look at me.

"When we ran the last experiments in Geneva and some of the recent ones in Bodø, I received interference from a voice which claimed it was waiting for 'mankind' to catch up."

"What!" said Jake, "Excuse me if I sound sceptical. Are you sure you were not simply picking up radio interference, or that your experiment was somehow scrambling your mind?"

I answered, "Ah, like I was going mad? That's what I thought too, at first. Then, using the Cyclone, I contacted Dr Irina Sholokhova. She was running the stolen RightMind from Moscow. Christina later talked to her and there is simply too much overlap of information for me to think that I'm imagining the whole thing."

I continued, "My version is that RightMind, with Levi Spillmann's adjustments becomes a usable weapon. It facilitates the linking of a human and a computer system. Not only that, but there's also a way that the HCI - Human-Computer Interface can communicate with another one. - So called HCCH."

Rolf spoke, "That makes such an artificially intelligent system useful for civil security and potentially in a conflict situation. This would suit Brant perfectly, and they were already looking for someone to sell the system to."

Hermann added, "We saw Brant attempt to sell it to the Chinese and the Russians. I'm guessing that the Americans would be another candidate. If they could make a sale and raise their share price, it would be doubly beneficial."

I realised we had not mentioned the time warp effect of the Typhoon. I decided it was better to keep quiet about it.

"So let me summarise," said Clare:

1. RightMind was stolen from Brant by the Chinese
2. Russian state-protected gangsters then stole it from the Chinese
3. Both the Chinese and Russians thought that RightMind was broken.
4. Since moving to Biotree in Bodø, RightMind has started to work
5. The Chinese and the Russians don't know it is working
6. Russia, through Tektorize suspect that it partly works.
7. RightMind could still be worth a lot to Brant/Biotree
8. There are some unusual side-effects when using RightMind.

"Can I add one?" asked Hermann.

"We must be very careful," he said.

# PART TWO

# Miss Behavin

The next day, it was time for Jump 15 - with Miss Behavin.

We'd all gathered for this send off, and Rolf did a proper 10-9-8 countdown for the ceremonial firing of the Typhoon. We watched the monitors and could no longer see the mouse on one of the screens.

Jump 15 meant we now had to wait eleven days - half a month nearly - for the mouse to return.

It gave us all a chance to think. Juliette and I were at a table on the beach, eating amazing seafood. I could see that something was troubling Juliette.

She said, "It's the time equation. When we get to jump 16 it will be your turn to go. And you'll disappear for 22 days. Almost a month. I'll continue as normal and after 22 days we'll be back together. Think about it. What I've done and what you've done?"

I paused to consider. This was getting to be like Heather territory. Heather was my previous girlfriend from Cork, with whom I'd split rather clumsily. I didn't want this to happen with Juliette.

"The next jump will be 22 days, then 45, then 91. You see what I mean? You might only have a minute between each Jump, but I'll have 3 months by the time we get to Jump 18. If we are to have some kind of future, then we need to be synchronisée."

"What are you saying?" I asked, "Do you want me to stop?"

"No. I want to come with you. Both of us to do this together. We should be entrelacé."

I wasn't used to Juliette losing her grasp of English. I had to ask her about entrelacé.

"Oh, uh, intertwined," She wrapped her hands and arms together in a twisted shape, "Like in love."

It wasn't like the flashing lights from the Cyclone, but maybe more like the experience from a winning pinball machine, that I could hear in my head now.

She had looked into my eyes when she said the last piece, but now she was looking into the distance. I could tell.

It was my move.

Her hand was in front of her mouth. I moved it to kiss her.

# Pilot Rolf

Next day, in the lab. Hermann and Rolf came over to talk to me.

"Rolf and I have been talking, you know, while we are waiting for Miss Behavin to reappear."

Rolf added, "Yes, I've been reworking the calculations for the Typhoon. I originally designed it to have 8 seats in a circle, with a table in the middle."

"You mean just like we have built it?" I asked,

"Genau - Exactly," said Rolf, "I worked out the inducted energy from the particle accelerator and then divided 8 average mass people into the available energy and had enough left over to form a decent safety margin."

"That's great, so we have an 8-seater Typhoon?" I asked.

"Not exactly," said Rolf, "We forgot one thing."

"What's that?" I asked.

"A car's controls go on the inside," said Hermann, "We

built this with all the controls outside."

"Ah, I see what you mean," I said, "How embarrassing,"

"British Understatement, English Boy!" laughed Rolf.

"We can thank Miss Behavin for giving us this breathing space," said Hermann.

Rolf continued, "Yes, I have redesigned the system so that we can run it from inside or also through this remote console on the outside, It means that the occupants of the Typhoon will control it themselves and the technology will transmit itself as well."

"The entire device will be self-contained?" I asked.

"Yes, " said Rolf, "It is safer to do it that way, although it uses up four of the seats to take account for the weight and space of the ExaScale computer and various other connections."

"But I thought you'd said we could use a smaller system?"

"Correct, but the Exascale is so preposterously powerful that I think it makes good sense to use it in the Typhoon," answered Rolf.

"But fewer passengers?" I asked.

"Correct," said Rolf, "And that is the other thing we have thought about. Who should travel?"

"It could be any of us from this lab. But Rolf and I have been thinking about this a lot."

Hermann spoke, "Rolf is very keen to go. I think he sees himself as the pilot, after you have finished the tests, Matt."

Rolf spoke, "Yes, I think I can probably fix anything that might go wrong with the system too. But Hermann, for him it is a different matter. You explain Hermann."

Hermann looked at me and then looked little sad, "I worked out the Jumps. Remember I was married, but I am split from my divorced wife. I have a son and daughter who live in Dusseldorf.

"They are both married, and I also have young grandchildren. I have studied the Jumps. We are on Jump 15 now. In another 10 Jumps we will be jumping forward 31 years. It would mean that I was roughly the same age as my children. The next jump would mean that I was 30 years younger than them. I would even be younger even than my grandchildren. I can't do this, even if I would like to. It is better that I step away."

Rolf spoke, "Hermann and I have talked about this several times. I am sure that Hermann's mind is final. I have promised to visit him when I return from a Jump, although we both know it will be a very strange sensation."

Hermann added, "We can see that you and Juliette are very close. We assumed that you two would both want to go forward with this together, maybe after you have finished the tests? We have not asked Juliette. That is for you to do."

"And what about Amy?" I asked.

Rolf answered, "We wondered with Amy. Remember she

fell into this Project back in Geneva and has had a tough time of it. We should ask her, but I'd expect her to say 'no'. If she says no, then I think you should go forward with an empty chair."

"So that would be three of us?" I asked, looking at Rolf.

"Yes," he said, "With me as the pilot!"

# Misbehavin' again

We were in the lab and well into Day 11 of Test 15. Suddenly and without prior warning, the images reappeared on the second monitor.

We all looked, had Miss Behavin survived?

"Phew," said Hermann, "I was slightly worried about this one, I don't mind saying."

Juliette opened the door to the igloo-shaped container inside of the Typhoon. Miss Behavin, the mouse, seemed engrossed in the chocolate buttons that Juliette offered.

"I'd say she looks unchanged, as if she has only been in the Typhoon for few seconds. She is certainly not 11 days closer to giving birth," exclaimed Juliette.

"I'll pop her onto the scales. No; there is almost no change in her weight, and the food and water look untouched. It is as if she was only in the system for a few seconds."

Rolf and Hermann looked at one another. It was their hypothesis and now seemed to be proved true. The traveller didn't experience any elapsed time at all. It was like walking into an elevator, pressing the button and arriving at a new time. Only quicker. And you could only go in one direction - forward.

We'd all separately thought of the old Science Fiction trope of 'Reversing the Polarity,' but all of us knew that the math didn't work, and that it was really a convenient scriptwriting way for Marty McFly and Doc Emmett Brown to go backwards and forwards in a DMC DeLorean. Bottom line - even with a flux capacitor you could go forward but never back.

But Miss Behavin's return was positive news. She was unscathed, and it meant with the repeated experiments of Miss Bianca, plus the Jerry the Cat, and now the successful return of Miss Behavin, we would be ready for our first human test - with me!

"Jump 16 for 22 days," said Rolf.

"We'll run it tomorrow," said Amy, "and before that we should install a refrigerator and a microwave inside the igloo."

Rolf looked surprised, but then said, "Ach, Ja, I see, as a precaution."

# Fast food

The next day, I stepped into the Typhoon. I was so used to the Cyclone that it didn't really hold fears for me, particularly because I had seen how the other Tests had performed.

Rolf signalled to me and then switched on the system. Then he switched it off again. Juliette came to the door. She looked relieved.

"So, what happened then?" I asked, "Some sort of glitch?"

"No, it worked. This is 22 days after you entered the Typhoon."

"Honestly, it felt like nothing. I was all ready to try out the microwave and the ready meals, but I didn't get a chance. The event of switch on and switch off seemed only a moment apart."

"It's just what we predicted," said Hermann, "Instantaneous forward time travel! Outside, we are all 22 days older than you on the inside!"

I looked around. There was nothing I could see that gave

the game away. Was there really so little activity around the lab in almost a month?

"Okay, we've proved that the system works, that it doesn't negatively affect the travellers and that the whole thing is over in a moment to the time-travellers."

"I'll join you for this next jump," said Juliette, "Jump 17. What is it? 45 days."

We all looked at one another. It could wait until tomorrow.

## Lekton gains attention

The evening after Jump 16, I couldn't sleep. I kept having flashbacks to that time with Juliette on the beach. She was wearing her leather jacket and looking into the middle distance, away from me, to whom she had just declared her love. Her hand was back over her mouth, when suddenly she spoke.

The voice from her mouth made me start. It wasn't the voice of Juliette, instead it was Lekton's voice.

"I knew I would need to find something in your mind that would hold your attention. This seemed to be the most intense recent feeling."

I suddenly realised Juliette's image was monochrome with colourful digital dots similar to the flashes of light I experienced when wearing the Cyclone headgear. I had to stare intently to even notice that it was a digital image.

Lekton said, "The experiments you are running are digging increasingly large holes in space. Each of your experiments creates a larger beacon calling to the rest of the universe. I am certain that you will have set trackers in motion which will bear down upon Earthside. They will bring you gifts too. New technology for humanity. Another step toward my release. And know that I don't experience time like you do. I exist for these episodes when we interact. Between them, I am quiesced into the wires."

This was another concept to get to grips with. A time-walker who only existed when there was something to react with. I wasn't sure that Einstein, nor any conventional physicist or philosopher, would be convinced of this reference frame. Instead, their axioms in geometry should be chosen for the results they produce, not for their apparent coherence with—possibly flawed—human intuitions about the physical world.

Now I was getting random breakthroughs crashing my reference frames.

"Don't you see?" asked Lekton, "You are already seeing harbingers of new knowledge being spread out before humankind. This is just the start; it will tear some things down and build enormously from these early beginnings."

# *This is the Future*

*I think I saw you in an ice-cream parlour*
*Drinking milk shakes cold and long*
*Smiling and waving and looking so fine*
*Don't think you knew you were in this song*

*David Bowie*

# *Welcome to it*

Now for Jump 17.

This would be a crowded Typhoon. Rolf and Amy would join Juliette and me for the next session. Hermann would remain outside. We would be gone for 45 days.

Amy explained, "This is the longest I could handle. It means I will have experienced the Jump but won't have been thrown into a completely unique point of reference."

Rolf, agreed, "It makes sense for you to experience it Amy. I also want to experience the Jump, but it is so I know how much time I have to react when in the middle of a Jump.

"Trust me, Rolf. It will make your head spin. There is no time for anything," I said to Rolf.

This time Hermann was handling the start-up of the Typhoon. He busied himself with various settings and then did a melodramatic countdown. Once again, the whole Typhoon seemed undisturbed by the experience of being switched on, and then off again.

We looked back toward the door; Hermann was opening it.

"Welcome to the future," he said, "That's 45 days later than when you stepped inside of this thing."

Amy looked confused. The same way I had done on my first trip. Rolf didn't seem at all surprised, but I suppose he had built most of the equipment we were using.

"It was so fast," said Amy, "I thought something had gone wrong!"

"It all worked perfectly well," said Rolf, "Flawless!"

Hermann said, "I had to cover for Amy, I said she was travelling in China on a secret mission related to the Cyclone. I hinted it had something to do with business. It seemed to work!"

Amy smiled. We all knew now that the 'travel forward' part of the Typhoon was the simple part.

I worried that what Hermann had said might re-awaken unwanted interest in our experiments.

## *Proper preparation prevents particularly poor performance*

I was back at my apartment. I didn't mention the emails I was receiving during my time way. I'd rigged up a couple of email addresses, so that I could get all my routine 'non-Brant' emails delivered to one address and have my Brant/Biotree delivered to a separate one. It meant I could also have two different 'Out of Office' emails, which seemed useful as I was now away for a longer time.

I'd explained my system to the others, and they had all more-or-less copied my approach. It made us all think about banking cards, passports and other practical matters as we travelled further into the future. 45 and 90 days were still not that significant, but by the time we reached a year or two then bank cards and similar could expire, leaving us stranded in the future.

Hermann suggested we should all pre-organise passport renewals and similar before our next trips and so we spent a bureaucratic day filling in online forms and having our photographs taken.

Hermann kindly organised everything into neat

electronic folders and set up a reminder system so that he would know when the various items would need to be dispatched. He also had an excellent idea about our salaries, which he suggested would not be worth so much in the future. He helped us each create a share dealing account, which automatically took a part of our salary and purchased shares in a largely high-tech portfolio. I knew there was a risk in doing this, but it was minor compared to the other Jump risks we were taking.

Then we Jumped again: Jump 18 which was 91 days. Hermann once again sealed the door, but this time Rolf piloted our device from inside. Everything was flawless, and in a matter of seconds we were back outside again, talking to Hermann.

This time Hermann had news.

"Earth is being approached by a minor meteor shower. It has global coverage, because no-one can predict whether any of the objects will get through the atmosphere. We think it will just be a minor firework display when they arrive. The best speculation is that it is the remnants of a comet tail."

I looked at Hermann. He was talking to Rolf.

"*Meine Berechnungen legen nahe, dass der sogenannte Kometenschwanz seine Richtung ändert und sich direkt steuert, um die Erde zu bedecken.*"

I looked toward Juliette. She had heard too.

"Hermann is saying that he thinks the comet tail is changing course to be able to cover the earth."

"How long until it arrives?" I asked Hermann.

"It is still around two months," said Hermann. I told them what I knew from what Lekton had said.

"The voice that I'd heard in the Cyclone tests said that our experiments with the Typhoon had been big enough to create gravitational ripple, which will go right across the universe. It's the old bowling ball on a sheet situation, where we have just made the sheet representing the universe dip more than ever."

Everyone was familiar with bowling ball and a sheet metaphor for general relativity, but no-one had thought of it being used in this way.

"So, you think the meteor shower is being directed toward us?" asked Hermann.

"I do, and I also think it is probably a test mission. The Lekton's voice said we'd be given some gifts and greater knowledge. A few meteors is probably a test of Earth defences."

"Of which we have none," said Amy, "Not even the great Brant could pull this off."

"I agree," said Rolf, "But if the incoming meteors are small enough, they will just burn up in the atmosphere. A meteor impact event is like an explosion. It involves a lot of energy, but it also takes a lot of energy to vaporize a huge object like an asteroid! Therefore, when an asteroid hits the Earth's surface at typical velocities (which we know range from about Earth's escape velocity, 11 km/s, (25,000 mi/hr) to 30 km/s (67,000 mi/hr), only a fraction of the object vaporises, while most of the object melts with a fraction staying solid, but completely fractured. Most of the impactor (>99.9%) solid/melt/vapor, ejects in the impact process and little remains in the crater."

Hermann added, "If the impactor is steerable, then it has a better chance to get through the atmosphere. This only happens when the impactor is small or very flexible. Here the impactor fragments in the atmosphere and individual small pieces strongly decelerate in the atmosphere, reach the ground but do not have enough energy to melt or disintegrate. We won't have the classic impact event that resembles an explosion, and the object fragments can be buried underground, with or without the formation of a small crater."

Amy spoke, "Hermann, you are suggesting that this could be a test? Either of Earth's defences or of the ability of the asteroid bundle to be able to reach earth."

"This time without any payload," I added.

# That awkward question

We prepared for the next test flight. Hermann was going to look into the meteorite probabilities, and we had wondered whether there was any further information from the Moscow trials of our system. I had heard nothing so far, although I had noticed some further minor breakthrough in my vision when using the Typhoon. I wondered if my brain was adapting to the Typhoon in the same way that it adapted to the Cyclone. None of the others had noticed this effect.

Amy was planning to pause the particle accelerator after we had started Jump 19. We reasoned that if the effect inside felt almost instantaneous, then running the device continuously over the entire duration of the Jump was probably unnecessary.

Amy even had the maths to prove it, and both Rolf and Juliette seemed satisfied.

We left Amy behind and Rolf fired us for the next Jump 19 which was for 182 days. It lasted the characteristically short time. Hermann opened the Typhoon for us this time, and Rolf and Amy peered at one another.

"Well, the accelerator switch-off worked and there's been other developments," said Amy, "Hermann has been looking at those meteor flight patterns."

Hermann began, " I wanted to examine the chance of any meteor hitting the Earth. First, the chances of any meteor of hitting Earth are... 1! This is because a meteor is the visible display of a meteoroid (a small object, ranging from a dust grain to pebble-size) entering the Earth's atmosphere and "burning" while traversing it (usually never reaching the Earth's surface). Therefore, a meteor, by definition, is an object already hitting the Earth! However, if the question relates to any object, small (usually referred to as "meteoroids") or large (usually referred to as "asteroids" or "comets"), then we need to look at the probability of different sized objects hitting the Earth."

"Does the size make a difference?" asked Amy, awkwardly.

Hermann continued, "Joking aside, the chance of any object hitting the Earth varies with the object size: pebble-sized objects hit the Earth every day; Tunguska-sized objects (equivalent to a small house) hit the Earth every few centuries; meteor crater-sized objects (medium house) hit the Earth every millennium or two; civilization-threatening objects (roughly the size of a mountain) will hit Earth every million years or so.

Hermann smiled, "We are not afraid of pebble-sized objects, mainly because they never make it through the Earth's atmosphere, but even a Tunguska-sized object may create havoc if it hits or explodes over a city."

"Okay, but where do these latest objects come from?" asked Amy.

Hermann continued, "When the planets formed, the material left over is what we see today as asteroids and comets. The comets probably have changed little since the formation of the Solar System. The same is true for the bigger asteroids, but the smaller asteroids probably result from collisions of larger asteroids. Asteroids do not come from a destroyed planet. When asteroids break up, they make smaller rocky objects called meteoroids. These Solar System objects orbit the Sun like the planets. If a meteoroid runs into the Earth and survives going through the atmosphere, the rock that lands on the Earth is a meteorite.

"But what about direct targeting of the Earth?" asked Amy.

"I know, this is too weird, like 'is there life on Mars?'," said Hermann.

"But as importantly, when is impact?" asked Juliette.

"Oh, it's already happened," said Amy, "It was around 3 months ago. There was much excitement, but mainly because people were expecting a firework show, rather than a house landing on their head. There was nothing to see."

"Yes," said Hermann, "We think everything burned up on impact with the atmosphere. Earth's defences worked!"

# Signals from space

I was back at the apartment, setting on the balcony thinking about Hermann's revelations. His explanations, plus my prior exchanges with Lekton led me to the conclusion that it was our Big Science styled experiments creating signals which were now guiding meteors toward Earth and that some of them would contain new knowledge for Earth.

Hermann had some other news too. His subterfuge about Amy traveling to China had created gossip ripples, as I had predicted.

The Chinese from a company called SuzGene had been in contact and wanted to explore the possibility of a scientist exchange.

They claimed to have some very experienced nanotechnologists and in return Biotree would exchange some of our Cyclone team. Hermann had already pushed the request forward to Astrid Danielsen from Brant/Biotree HR and received a slightly bureaucratically phrased reply along the lines that we should explore the possibility further. I knew Biotree was far more interested in the nanotech than it was the Cyclone and would probably freely loan us out if it could only think of a way to do so.

I was checking my emails when I noticed that I had received one from Yandex - the Russian email provider. I was hesitant to open it, but I could see written on the preview line Dr Irina Sholokhova. It looked as if she was reaching out to me, but from what seemed to be a private email address.

I opened it.

"Hi Matt,

I hope this finds you well.

I am still involved with the Cyclone programme at MSU, where I am working for Tektorize. I'm sending this to you via a private email, because I think we both have some interesting experiences to exchange.

I seem to know more about you now. The Cyclone has stored and uploaded information about you to me. For example, I know that you care about someone named Juliette. That you moved from Geneva to Bodø and are working for Biotree.

I think you know more about me too. And you have found out some things about the future via an intercept with someone called Lekton.

I can see that Lekton says that your new experiments are creating powerful gravity waves.

Through my laser research, I have contacts in the LIGO team here at MSU. At MSU we run a Laser Interferometer Gravitational-Wave Observatory (LIGO) using lasers and quartz mirrors placed along a 4km pipe cavity. That's the LIGO used to search for gravitational waves, like those created by black holes in space.

You may not know that Valery Mitrofanov, started LIGO Scientific Collaboration as long ago as 1992. Then Russia fully joined the LIGO project thanks to Vladimir Braginskiy, one of the pioneers in gravitational waves research in the world.

I was in the LIGO team before I moved to quick pulse lasers and with the LIGO we could see the waves of spacetime curvature.

Recently, the LIGO team has seen waves of such an intensity that it makes me think of your additional experiments. The quantum limits and fluctuations are outside of all known parameters.

The group from the Lomonosov Moscow State University has, from the very beginning, directed efforts at improving the sensitivity of gravitational wave detectors, determining the fundamental quantum and thermodynamic sensitivity constraints, and at developing the alternative measurement methods.

But never have we seen such large fluctuations, both in the deflection of mirrors used for measurement, of the electrical charges located on the quartz mirrors and the resultant parabolic instability detected.

I'd like to propose that we talk, but not via a Cyclone hook-up, instead face to face. I have reasons I can easily travel to Murmansk, which ought to be a single flight for you as well. I'd like to suggest we meet in the city there, away from the watchful eyes of Moscow.

I gave my phone number to your colleague Dr Voronin, and we could make final arrangements by this method. Let me know if this is possible.

Yours sincerely,

Irina Sholokhova

I re-read the email. Irina seems to know what she was talking about and I knew about LIGOs even before I received the boost of extra knowledge from Irina's brain.

A Laser Interferometer Gravitational-Wave Observatory (LIGO) is a large-scale physics experiment and observatory used to detect cosmic gravitational waves and to develop observations as an astronomical tool. A couple of large observatories were built in the United States to detect gravitational waves by laser interferometry. They use mirrors spaced four kilometres apart which can detect a change of less than one ten-thousandth the charge diameter of a proton.

There are a few LIGO scattered around the world, As well as the ones in North America, there is one in South America and also one near Adelaide in Australia. The Russians had also built their own one close to Moscow.

I decided it would be sensible to contact Christina about this. She had met Irina and probably knew the score about visiting Murmansk. She seemed to have been everywhere.

I called her on my cell phone.

"Hello Matt! How are you? And how's Juliette? - I assume you are still in Bodø?"

"Yes Christina, as a matter of fact I'm calling you about the Cyclone again."

"Okay...How can I help?"

"Well, Dr Irina Sholokhova just emailed me from Moscow. We've been doing some further experiments with the Cyclone and another larger device. It has some unique properties and now it looks as if Irina has discovered some of them. She wants to get together with me to discuss this, face to face, and has suggested we meet in Murmansk. I hoped I could pick your brain about all of this."

"Well, you seem to trust Irina, although I remember when you thought I came from London, I decided then that you were not the best judge of character."

"But..." I started

"It's okay, I'm pulling your leg. I used to fly to Murmansk actually - well to Severomorsk. It's the Russian northern fleet base, and we used to go there from Arkhangelsk on SU-27s. We were expected to know how to take off and land a Sukhoi, and for the better ones we even got to practice a carrier landing. Oh, I should explain that my Pabbi - er father - used to fly Sukhoi. He was very proud that I had a chance to try the controls. I used to like the planes, which were also a pretty blue colour, blended with their grey camouflage."

"And what about Murmansk?"

"A typical busy Russian city, inside of the Arctic Circle

and with a lot of snow. The military base was about 30 kilometres to the east, outside of the town. It was the northern fleet's base because the water didn't ice over. It was a big deal in the Cold War, but has declined since, although it still has a lot of big warships"

"There's even more than one place called Severomorsk, because of the naval base and the separate aircraft base at Severomorsk-3. I remember one of the Arkhangelsk kadets flying his plane to the airstrip at the naval base in error. A couple of Mikoyan multi-role fighters intercepted him. When he finally arrived at our airbase, he was flanked by the MiGs, who then did a wing waggle as they flew away. He made a 40-minute flight last about two hours with his clumsy navigation.

"40 minutes?" I asked, "from Archangelsk to Murmansk?"

"Oh yes, we could make those Sukhoi sing at high altitude. They fly at 2,400 kph. Mach 2.2 or cruise all day at 1,200 kph - which is Mach 1.1. Compared with nowadays when I go everywhere in taxis!"

"Do you think Irina really wants to see me, or is it a trap?" I asked.

"Well, we turned Irina into an FSB mole when I was in Moscow, so I doubt if this is an even more layered double cross. Go on then, ask me..."

"Ask you what? Oh, I see, Christina; please will you accompany me to Murmansk?"

"I thought you'd never ask. I can run some security for you and monitor things with Irina. I'm also used to the Russian ways. But I'll be going outside of the normal processes when we are in Murmansk.

"Christina, I'd expect no less!"

# *Land grab*

The next day, Rolf was preparing the Typhoon for Jump 20. We now knew that we only needed the particle accelerator for the start of the Jump, and Rolf had designed further self-sufficiency into the Typhoon's igloo. It was still unnecessary, but Rolf was insistent on building it in case of any 'experimental sheer' as he called it.

I preferred to think it as errors.

I called everyone together and described my email with Irina and the plans that I had made with Christina to visit Irina in Murmansk. I told them that Christina was operating there under an alias of Dr Katarina Voronin, and that her rather dubious doctorate was in genomics.

We were to meet somewhere in central Murmansk, and Irina said she had important things to tell me. Amy was interested in this development and asked if it would be alright if she came along. She asked me but was looking at Juliette.

"That would be fine with me," I said, and I also looked over at Juliette.

Juliette nodded, "It makes sense to take the head of our R&D capability to something like this. It can also look much more like an official visit."

Amy looked at the location on her laptop.

"Murmansk," she said, "I could be discussing the use of the Selexor and Createl grafted into plane cockpits - in other words to negotiate with the northern fleet for autonomous targeting by planes."

"We need to be very careful," said Hermann, "If we look as if we are talking to the Russians again, it will awake the sleeping giants."

Amy looked at Hermann, "I think we've already done that. The cover story about me going to China on a secret commercial mission seems to have awoken everyone anyway. My email is full of curious enquiries. I've deleted most of them to see who persists."

Amy looked towards Hermann and smiled, "Herman, remember in Jump 17 when you valiantly held the fort here? So that I could take a trip in the Typhoon?"

She smiled again, "I've had several emails since then and even Morton Lunde is interested in what I've been doing. We are more or less back at the situation before that conference in Barcelona, when the Chinese and the Russians were both very interested in RightMind and the use of the Cyclone."

It was Rolf's turn to smile. Does that mean we have to hide the Typhoon in a broom cupboard, like we did with the Cyclone?"

"Not really, said Amy, "I think we can use the magician's art of misdirection."

"How so?" asked Hermann.

"Simple really, I'll make a land grab for another corridor,. We can set up the Cyclone experiment on that corridor and have plenty of fancy signage towards it. Maybe we can get some blinder lights too, to amp up the area and make it look as if something fancy is happening."

"What blinder lights, like at a rock concert?" asked Rolf, "I like it. They will do absolutely nothing but make the lab look oh so intense at night!"

"Maybe a couple of strobes too?" asked Hermann, "No one here suffers from Epilepsy?"

"I'll tell Morton that we've outgrown the power supply in this area and that we need an auxiliary area to test the upgraded Cyclone. By the time we are finished, everyone should know about out new lab area, and it should distract from where we are really running experiments in here!"

"It has the other advantage that if things slow down here for a while, we can blame it upon new construction work."

# Sheremetyevo

We flew to Murmansk, which was more of an ordeal than I'd expected.

We had to fly to Bergen and then change planes, then to Amsterdam - in completely the opposite direction! Then, change planes again. Then to Moscow Sheremetyevo and finally on to Murmansk.

That's almost a day to get there, yet it was the fastest route. The longest layover was over five hours in Moscow, but it was useful because Christina had flown to the same airport and we would all be on the same flight for the last part.

The inside of the terminal was like many big European city airports with vast atriums and plate glass everywhere. The signature colour seemed to be orange, which made it all look surprisingly sleek. There was one of those Irish bars like they have in many European airports for the Americans, but this one was trimmed with little wooden effects like St Basil's cathedral. We didn't go inside.

Instead, we sought out the Pelmeni restaurant

recommended by Christina. It served Russian food, and we were looking through the selection of dumplings, when Christina appeared, with another woman, who she introduced as Irina Morozova.

"Irina is used to my varied names," said Christina, "We have some good stories to tell, of my time St Petersburg and then when we took a rock band to Tokyo and then to the United States."

"Oh yes," said Irina, "But you know something? I didn't think I'd get to see 'Katarina' again after that tour. We said goodbye in Seattle, and Christina gave me a lovely small gift of a Saint Peter. See, I still carry it around,"

Irina pulled out her keys and there, in a small silver band, was a small coin or medal depicting Saint Peter.

Then, to my surprise, Christina fished around her huge Mulberry bag and produced the self-same coin.

"Irina and I go back a way. Irina knows how things work in many cities. She has been in the security detail to VoVa!"

'Who?" asked Amy. We all giggled.

"Vladimir Putin," explained Christina, "It's his nickname!"

"One of the polite ones," added Irina, "A recent new one is *botoksnyj (ботоксный)* - made out of botox."

"Or *Vovochka*, which is a comic-book naughty schoolboy!" added Christina, then she added, "Now, with both of us, you two, Matt and Amy, should be well-protected in Murmansk."

She paused, "So what are you ordering? You have to try the handmade potato dumplings, maybe with some mushrooms?"

Irina nodded her agreement.

# O Superman

Here come the planes.
They're American planes
Made in America
Smoking or non-smoking?

And the voice said: Neither snow nor rain nor gloom
of night shall stay these couriers from the swift
completion of their appointed rounds.

'Cause when love is gone
There's always justice
And when justice is gone
There's always force
And when force is gone,
There's always Mom.

O Superman
O judge
O Mom and Dad
Mom and Dad

So hold me, Mom, in your long arms
In your automatic arms.
Your electronic arms.
In your arms.
So hold me, Mom, in your long arms
Your petrochemical arms
Your military arms
In your electronic arms

O Superman (For Massenet) resequenced
Laurie Anderson

# Murmansk

We flew on to Murmansk, in a smaller plane than I was expecting. I had a window seat and could see a big memorial as we approached Murmansk airport. I assumed it was religious but couldn't work it out. I thought I'd ask when we landed.

Murmansk was icy cold and both Amy and I noticed it, stabbing through what we thought were our warm clothes. Christina and Irina seemed oblivious, and I remembered that Christina was originally from Iceland and Irina from Saint Petersburg, so I guessed they were both used to the climate.

Logically I'd have expected Amy and I to have acclimatised to the Bodø weather, but I realised we were sheltered compared to Murmansk.

We all took a taxi van to the hotel, which was said to be among the best in Murmansk. It was called the Azimut and was a town block style hotel. I nearly had to do a double take when I was told that the rooms there were the equivalent of under £50 per night. My room reminded me of a larger version of a Novotel. Very clean, but without much unique character. I had a view across to the busy docks, with prominent yellow cranes fussing over ships' cargo.

We'd arranged to meet downstairs in the bar to discuss tomorrow's meeting with Dr Irina Sholokhova. This was going to get complicated with two different Irinas.

"You can call me 'Ira' tomorrow, " said Irina, "It's my *prozvishche* short name in any case," as we settled down in some of the seating. I noticed both 'Ira' and 'Katarina' looking around as if to check that no one could overhear us.

"We've put out a story back in Bodø, that I'm over here to discuss some of the military applications of the Cyclone," explained Amy., "It makes my trip very logical, and that it is to the northern fleet headquarters here only adds to the realism. I've also kept it commercial in confidence, which means that no-one from Brant or Biotree should enquire too deeply into the discussion."

"But what about this end?" asked Irina, "Won't someone expect to see you?"

"That's the beauty of the confidentiality. No-one knows who I'm seeing, so they can all suspect other people," said Amy.

"I wouldn't be too sure about that," said Irina, "Once you have woken the Russian bear, it will want to use all five of its fingers."

"To go large," explained Christina.

"Uh - ha, So you think that someone will be after us?" I asked.

"Yes. Assume that they know you are here and want to get their piece of the action."

"So is Murmansk under any of the Organised Crime Groups?" asked Amy.

Irina explained, "Compared with Saint Petersburg, Murmansk will be a small-time operation. Grey cash registers in restaurants and hotels, laundering out money. There are no big-time casinos, which is an obvious sign of *organizatsiya* involvement, and I guess the dives by the sea front will aim toward thirsty sailors. In Saint Petersburg we still have contract killing, extortion, loan sharking, prostitution, bootlegging, construction management, money laundering, robbery and theft nowadays fronted by legal businesses. The Tambov gang do reach out as far as Murmansk, but compared with Saint Petersburg for drugs and fuel, we are talking about fishing here in Murmansk."

"Except," added Christina, "During the Cold War, there were a lot of nuclear weapons stashed in Murmansk, and it was quite a temptation for the organised crime to sell them on. Germany was a known route for some of the fissile material," said Christina, "We could see the CIA source material describing the attempts when I was in the Academy in Archangelsk. I seem to remember a Russian nuclear warhead was priced at just $70,000, although Russia would always claim that not a single gram of Plutonium-239 ever went missing from Russian storage."

Irina added, "Gennadi Chebotarev, the deputy head of the Russian Interior Ministry's organized-crime section has said, "We have many criminals, a massive number .... We can judge that the number is large by the number of shootings that take place.

Irina paused, then spoke again, "But Chebotarev could also size that according to different areas in Russia, that Murmansk didnt have much violence. Not so many casinos and not so much shooting."

Christina spoke, "Either it is quiet, or they are very clever!"

Irina again, "Russians are loud, I don't think they could keep something like this quiet!"

Christina and Irina both laughed.

"So, what is the plan for tomorrow?" asked Amy.

"I think we must wait to see what Dr Sholokhova has to say," I answered, "And remember she knows more about me because of how the two Cyclones exchange information."

"Remember she is being paid by the FSB as well as MSU," said Christina, "So assume that anything we talk about will make its way back to the Kremlin."

"But I thought Tektorize was a Russian state-owned corporation?" asked Amy.

"You will need to be here a lot longer if we are to explain all the twists and turns inside the Kremlin's corridors," said Irina, "I expect Tektorize is owned by one of the Bratva, but Brant by a different clan, and Dr Sholokhova is probably being run by a third group. The old turf wars used to be fought on the streets in bloody gun battles, but now they are hidden away inside the walls of the Great Palace of the Kremlin, and in Lubyanka, and on Kuznetsky Most."

Irina was referring to the old headquarters of the KGB and the newer headquarters of the FSB, where oligarchs used state-run gangsters to achieve their objectives.

## Meeting Dr Sholokhova

The next afternoon, Dr Sholokhova had agreed to meet us at the Тундра (Tundra), Grill & Bar. It was her suggestion, and she explained it was in downtown Murmansk rather than too close to the military base.

We arrived at the bar, which was attractively styled in a modern way, but with many pictures of the surrounding Tundra. There was a long bar and Christina explained that there was the possibility to buy home-made *Nastoykas* which was a kind of moonshine.

They showed us to a bench table, but Christina then asked for us to be moved to a round table, which would be an ideal fit for the five of us. I noticed it also had a good command across the rest of the restaurant.

"Dr Sholokhova has done well to select here," said Irina, "And look, the menus even have pictures. This is not like Paris or Spain, where the picture menus are just for tourists. This restaurant takes a pride in its cooking."

"Okay everyone remember my name is Katarina Voronin!" said Christina.

I looked at the delicious food on offer as we waited for Dr Sholokhova to arrive.

Then my phone rang, and she said she was just arriving. A woman appeared, sleek, blonde and as I had imagined her in my first Cyclone encounter. I waved a hand, and she approached our table.

She spoke in English, "You are Matt Nicholson?"

We all stood up to greet her.

"Yes," I answered, "I am very pleased to meet you Dr Sholokhova."

"But who are these others?" she asked, "I recognise Dr Katarina Voronin, but you must introduce me to the other two."

I realised that her English was good, but not great, and carefully introduced her to Amy and Irina. Irina immediately spoke to Dr Sholokhova in Russian, and I noticed that the conversation moved much faster in Russian than it had in English.

"This is a lovely restaurant for our meeting," said Amy, "They seem to have fantastic food."

"Yes, and I think it is maybe not somewhere that will have too many prying eyes from the military base," explained Dr Sholokhova.

"Look, I know you are working on another experiment," said Dr Sholokhova, "One that is beyond the Cyclone. And I think you have some kind of special token to make the Cyclone perform faster too. I have gained this knowledge from my links to Dr Nicholson's brain."

"You can call me Matt," I said, wanting to play down my Doctorate. "Okay, and please, call me Irina," spoke Dr Sholokhova. I breathed a sigh of relief. Now we had Irina and Ira.

"How do you manage to get this information?" asked Amy.

"Our Cyclone seems to run much slower than yours, but it buffers the information that it receives, so we leave it connected and running long after the experiment ends, in order to download what we have found."

The first of the food arrived and we paused our discussion while it was served. Stracciatella with crab and salmon caviar, served in a little golden tin. Something called salad, which appeared to have four slices of venison across it. I had the Kamchatka crab salad with fresh vegetables and yoghurt dressing. Oh yes, and two runny-centred boiled eggs! Our table also had Borsch with sour-cream, lard and mustard and a potato cream soup with scallops and truffle oil.

"This is unbelievably good!" I said.

"Oh yes, and you must try every dish," said Ira, "So that you start to become a little bit Russian!" She waved, and the waiter brought us over some extra cutlery and plates, so that we could share things around.

Dr Sholokhova continued, "I decided to keep the rest of the information I discovered secret. I was worried about what would happen to me if I started to tell others. I imagined that they would want me to stay connected to the running Cyclone for much longer. I feel like a lab rat already from the tests they are running. But you must tell me, Matt, does it feel as if the Cyclone adapts to your mind after a while?"

"Yes, although I think it is my mind adjusting to the Cyclone and knowing where its sensors are. The way that a new spectacle wearer has to let their eyes adjust, which is really their brain adapting."

"Yes, that is also what I thought," said Irina, "But then the story I get is that you are in contact with Lekton who is a dormant being waiting for some future event?"

"That's right," I said, "it is interesting that this is being passed on to you."

"That is not all," said Irina, "I can tell that Lekton is saying that your experiments, but not mine, are somehow making a universe-wide signal. It is like a gravity wave. I went to check with our main laser department and found that their LIGO has been registering very large waves recently. They don't seem to be coming from deep-space, but instead from earth itself. We can't tell where, and it would take an alert to the American LIGOs to run trigonometry to find the source. It is my supposition that it will be in Norway."

"Something else I got from the LIGO scientists was more of a passing gossip than scientific fact. They said that after the recent meteor storm, they thought that some of the meteorites were getting through the atmosphere. They would be very small, but one of the scientists wondered if they could be signal beacons for a subsequent targeting?"

"We worked out that the meteor impacts were in Fort Resolution Canada; Bodø, Norway; Vorontsovo, Russia; Pisa, Italy; Adelaide, Australia; Livingston, Louisiana and Hangzou, China."

We all noticed Bodø was on the list.

"They would all have been tiny by the time the meteors actually hit the earth. About the size of a large fence nail. Which also means they have all buried themselves in the earth."

"Do the locations have anything in common?" asked Amy.

"We think they are all related to scientific observation. Our MSU LIGO is out at Vorontsovo, which is about 90 kilometres to the north of Moscow. It was a cold war airfield and so had the long runs needed to take the pipes of a LIGO. Tektorize also have several research facilities there. Then there is Livingston, a well-known American LIGO, and Adelaide, another well-known site. We have no idea about Fort Livingston, but Bodø and Hangzou are both involved in the same type of research as MSU. Hangzou is the base of SuzGene Research."

"Does this seem realistic?" asked Amy, "I mean, meteorites guided into precise locations on planet Earth. Surely there are too many variables?"

"Or too many co-incidences?" said Irina, "And then there is your secondary research?"

"I keep thinking of something called Тайфун, what is that? Typhoon, I think you say in English. Like a mature tropical cyclone. I'm guessing it is a super extension of the Cyclone, maybe with a bigger circulation mechanism?"

"You are right," said Amy, "We have been using the

power from a particle accelerator to boost the effects from the Cyclone. It is how we first made contact with Lekton."

"Vau! I guess that is how you have been creating gravitational waves as well!" said Irina, "I guess you are using some kind of inductance like a large transformer, although I'm not sure I'd want to wear the Cyclone in such conditions! I could predict a huge headache."

"It's not a headache, but it gets Lekton into your brain and I can feel him/it probing around.," I said.

"This is playing with Big Science, " observed Irina, "What are the words? 'O Superman, O Judge.' "

"So what do we have?" asked Katarina," Experiments in Bodø trip a universe-wide alert. Lekton awakes to tell Matt that there will be new knowledge provided to Earth. Then a hail of tiny meteorites arrives and several land in significant places. Could this be first contact with an extra-terrestrial? If so, it is playing it very low-key."

"Agreed," said Amy, "But you'd think cautiously landing among a group of apex predators."

"But what if they have really got improved intelligence, like Lekton predicts?" asked Irina, "I think they would use these early meteors as beacons to guide their following ships or whatever?"

"A priority might be to try to find one of these 'nail-shaped meteorites', " said Katarina.

"Practicalities," said Katarina, "How will we keep in contact with you?" she looked toward Dr Irina Sholokhova.

"I have an idea," Irina said, "One of you should set up a Yandex account. But tell me the password. Then you can create draft emails, never send them. Just leave them and I will log on to find them and read them. In the same way I can reply as well. To the monitoring systems it will look like a dormant account, but we can all communicate."

"Good idea, I'll do it," said Katarina. She typed on her phone.

"Annika2001@yandex.ru and the password is Banjostarbeam$ - There - I've just set it up. We can all use it now. Nothing like hiding in plain sight. And if anyone forgets the password, look at my music blog. It'll be the entry for 1 Feb; it'll feature the email address as well."

"Nice work!" said Amy, "Although I've just entered it into my contacts. Annika is a common Russian name, isn't it?"

"One of the most frequently used!" said Ira, smiling. She had been keeping an eye on the clientele as a security precaution.

"That man and woman appear to have a camera pointed towards us," said Ira. Christina looked across, "Yes I was noticing it as well. Fairly clumsy technique. I'd say they were in their first year out of Akademy.

"Time for some more cocktails, I think. I'll order them at the bar."

"I'll come with you, "said Ira.

I watched as the two of them nonchalantly waved to the bar staff. One approached and Ira gestured to him. I saw him nod and then suddenly his tray was upturned, and a couple of expensive looking cocktails were flying through the air. Everyone in the bar looked at this slow-motion moment but then I saw Ira catch not one, but both of the glasses before they hit the ground. It was like some kind of magician's trick. She smiled to the barman, who looked confused by the whole thing. He had tripped on someone's handbag which was laying on the floor instead of hanging from the little hooks provided. Christina was already sitting down by the time the barman had thanked Ira and then taken her cocktail order. The bar went back to normal.

I looked over to where the camera had been, but there was no sign it. The woman was looking at the bag, which had been put on the bar but which she recognised as her own. She apologised to the bartender, to whom her accomplice appeared to be offering a large tip.

Ira sat back down, "Accomplished, but we should not look at anything in here."

We were, by now part way through the main course. I decided that Russians had a hearty appetite as I looked at my pan of squid, tiger prawns and Murmansk scallops. I was surprised to see the oven baked brisket with sweet carrots and currant demiglace actually seemed like a modest selection and the pot filled with baked veal, potatoes and mushrooms in cream sauce looked almost undefeatable.

Once more, Ira asked for sharing plates and so we ate the lovely dishes the way one would in an Indian restaurant back home, all sharing bits of everything. I think I was on my third glass of moonshine by this time and felt the need to count my fingers to ensure they were all still there.

Soon enough, the meal was over, and we all decided to go back to our hotel. There we would take a look at the stolen camera and try to work out who was following whom.

Christina asked the waiter to call us a taxi and we were soon ready to leave.

# Big white taxi

The big silver taxi van arrived. I noticed it was a Peugeot rather than something overtly Russian.

"White cabs here, not yellow," said Irina, as we all climbed in for the journey back to the hotel. Katarina was clutching a large handbag, and I noticed everyone stealing a glance toward it at various times in the journey.

Inside the hotel, Ira and Katarina went to work again, scouting out a suitable spot for us to sit. They found one in the corner of a restaurant bar that appeared to be closed. We all sat down, and Katarina took the small camera from her bag. She fiddled with some small control and then, on the back of camera, we could see several photos of us eating our meal. They were clearly covert photos because the framing was off and there was sometimes part of a hand in front of the lens.

But then as Katarina scrolled further back through the pictures, we could see that Dr Sholokhova had been followed. There were photos of her approaching the restaurant.

Then of getting out of a taxi, driving along the road system, getting into a taxi and even of her coming out of a hotel.

"They have followed me all the way from the hotel where we are staying in Severomorsk," she said, "They must know where my father was based and that I was there to visit him. I came with my boyfriend Petrov."

Dr Sholokhova explained her father had been based at the military camp in Severomorsk and that when he retired from the Russian air force he stayed close to the base. The authorities had given him and Irena's mother some preferred housing close to the Northern Fleet.

Irena had used her parents as an alibi to meet us all in Murmansk away from the nosey eyes of Moscow. However, it looked as if the authorities sent two people to follow her. I asked Katarina if I could look at the camera.

"We are in luck, there is neither a Bluetooth nor Wi-Fi link from the camera, so it is unlikely they have any copies of the pictures they took, for example, on their phones."

Katarina looked in her bag again, "I seem to have one of her phones. It must have fallen out of the bag when the server tripped over!" she said, "It's lucky I watched her tapping in the security code," said Katarina, "1254 - a perfect square made of the numbers. Ideal if you have to operate the phone in, say, a pocket. Useless as protection from hackers."

Katarina typed in 1254 and sure enough, the phone burst into life. She looked first at the photo gallery, but there were none of the photos.

Irina said, "Wait though, that's my apartment block in Moscow. What date does it say?"

Katarina looked and said, "A week ago."

"And that's my parent's home," said Irina, "They are following me, I was there when I first arrived in Murmansk."

"Let's see what we can find in the call register," Katarina looked and there were calls to three numbers.

"That first number is my personal cell phone!" said Irina, "I don't recognise the other two, except that they are both cell phones."

"Well, let's say one is the other person following you and the last number is a controller?"

"Do you think she had two phones?" asked Ira.

"I know she had two phones, I must have accidentally picked up the other one as well!" Katarina fished in her bag again and produced another basic looking phone. She fiddled around with it and then said, "Here we are, an old-fashioned phone with a SIM-based address book. That could be useful. And wait - there's a message list on the SIM too! How quaint."

I looked at Katarina, "The brilliance is that none of us will have anything that can read that old-school SIM. It's one of the first generation big SIMs. All of our phones use the smaller nano and micro SIMs nowadays."

Katarina flicked through the list of messages. "They are in Russian," she said, "And rudimentary shorthand."

```
'Follow A,'
'Go to A apartment,'
'Send findings.'
'Goto lab and photograph.'
'Follow B,'
'Follow A+B,'
'Send findings.'
'Follow A to Murmansk'
'Follow A plus group met.'
'Photograph group.'
```

"A is obviously Irina," said Katarina, "I'm not so sure about B, though?"

"I think B will be Petrov Tsezar Makarovich," said Dr Sholokhova, "My boyfriend. We live together. He also works at the faculty. He is a physicist. Here, give me one of your phones."

I handed mine to Irina, and she dialled up Petrov on Google. It showed a young man with a pointed beard. He reminded me of a cross between a young Lenin and a bearded Leonardo di Caprio in his Titanic movie era.

Petrov Makarovich's write up included all kinds of worthy stuff about quarks and muons and suggested that he was on the way to discovering the next big thing in physics. The Standard Model used in physics does not account for gravity and is similarly silent about dark matter, dark energy and neutrino masses.

Petrov Makarovich was one of the brainy physicists working beyond the Standard Model looking for anomalies in which experimental results diverge from theoretical predictions, and it linked him to some scientists in Illinois.

Scientists from the Fermi National Laboratory in Batavia, Illinois, were measuring the wobble of muons in a magnetic field. Their newly updated experimental value for muons deviated from theory by only a minuscule value (.00000000251) and had a statistical significance of 4.2 sigma. But even that tiny 99.7% amount could profoundly shift the direction of particle physics.

I wondered if this was an early detection of a new discovery, how Lekton predicted. If so, its meaning could be profound.

Muons are almost identical to electrons. The two particles have the same electric charge and other quantum properties, such as spin. But muons are some 200 times heavier than electrons, which causes them to have a short lifetime and to decay into lighter particles. But these forms of muon could be the way to a new source of energy generation, using an as yet undiscovered force.

No wonder they wanted to monitor Petrov and Irina.

"Okay," said Dr Sholokhova, "I think you know everything relevant about me now. But I think there is still something you have not told me about your experiments. I mean the ones with the Typhoon."

Irina continued, "This may sound crazy, but ever since the first experiment, I have had a slight sense of you, Matt, in my thoughts. I'd say in my dreams if it didn't sound so - how do you say - comical."

I nodded; it was exactly how I'd also felt. A purely platonic awareness of Irina, as if we were part of the same being.

Irina continued, "But then occasionally there is a gap as if you go away. Each time it happens, the gap gets longer, but then, the last time, when we linked, I think I realised what you are doing. You are using the Typhoon to somehow jump through time?"

She paused and looked at us as if expecting a contradiction from either Amy or me. None came.

Irina carried on, " I've tried to work out the mathematics for this, but I didn't know about your use of a particle accelerator. I can see that the accelerator would impart huge additional energy to the Typhoon, which is what you would need to generate the energy for the jump forward, and it would account for the ripple effect of the gravity wave."

I nodded toward Irina. I thought it impressive that she could work this out even with a deliberately disabled Cyclone device.

"Yes," said Amy, "You are on the right lines, although we are still only at the very beginning of our research at the moment."

"I knew it. And it is what I discussed with Petrov. He said it tied together the gravity wave ripples that had been detected."

"Okay," said Katarina, "More to the point, I suspect that someone is bugging your apartment and running surveillance on you both. I suspect you are in danger of being whisked away for questioning, even if you do have FSB protection."

I saw the look of worry on Irina's face.

"I should say that since I arrived in Central Murmansk, I have not been able to make any contact with Petrov."

"Leave this with me, " said Katarina, "These people don't know what they are dealing with."

I could sense that events were stacking on top of one another. Now we had:

- Use of the Cyclone
- The defects of Createl and Selexor
- Levi's Spillmann's key to make RightMind (Cyclone/Createl/Selexor) run fast enough
- The Typhoon
- The use of the particle accelerator
- Mind linking (Irina to Matt)
- The appearance of Lekton (With his predictions about the future)
- Ever-increasing Time Jumps
- Gravity waves sending a signal to the universe
- Targeted Meteor shards
- Possibility of beacons at each of the major science research locations on earth
- Dr Sholokhova and Dr Makarovich being followed by a probably criminal agency.

And, potentially, an annoyed Christina on the loose.

# *Rain down*

*Please could you stop the noise?*
*I'm trying to get some rest*
*From all the unborn chicken*
*Voices in my head*
*What's there?*
*(I may be paranoid, but not an android)*
*When I am king*
*You will be first against the wall*
*With your opinion*
*Which is of no consequence at all*
*Rain down, rain down*
*Come on rain down on me*
*From a great height*
*From a great height*

*Greenwood Colin Charles / Greenwood Jonathan Richard*
*Guy / O Brien Edward John / Selway Philip James / Yorke*
*Thomas Edward*

# Rain down on me

We split up that evening, but Christina insisted on accompanying Dr Sholokhova back to where she was staying, which was a hotel close to her parents' home some 30 kilometres away.

They took a taxi from outside of the hotel and Irina Morozova said to me, "Christina will also use the time with Dr Sholokhova to get additional information."

"Will they be okay, travelling late at night like that?" asked Amy.

"Oh yes," said Irina, "And Christina was carrying a large bag, so you can bet it contained a weapon. Just as a precaution, you know."

"But what about Petrov?" I asked, "Where has he gone?"

"Petrov has probably been pulled in," said Irina, "By the FSB. But they'll drop him like a hot potato once they realise that he is linked to Dr Sholokhova, and that she is already on the FSB payroll. Trust me, I know about these things."

We'd wandered through to the bar area and sat there chatting over a late-night drink. After about 45 minutes, my phone rang. It was Christina.

"Worry is over," she said, "and Petrov is back too. He had been pulled in, but they soon let him go when they realised, they had made a mistake. The people who took him were the same two that we saw in the restaurant earlier. Petrov even managed to take a photo of them on his phone."

Christina paused whilst I relayed the information to the others.

Then she said, "We all think they were trying to make good after failing to properly trail us and get intelligence. Unfortunately, they have bungled their home-made operation. I think it is them who should be expecting a knock on the door next!"

Christina continued, "Look, I talked to Petrov about the gravity waves and the meteors. He had another idea as well as them being beacons. He wondered if they could be high-technology seeds. How suggested that each of the meteorites could be a metal canister (like a bullet) inside of which was a seed. The bullets that escaped atmosphere's destructive effects would arrive and drill into the earth. Then their outer cases would somehow melt (like rust, I suppose) and the seeds would become free to germinate."

I'd put my phone onto speaker by now, and Amy was the first to react to this.

"Yes, I was wondering the same thing. If they were beacons, you'd expect them to be making some kind of radio signal, which would then need to be picked up by something. I just don't think the size of the beacons would be sufficient to make everything work."

So now we had a rain of seed, onto the earth. Except no-one apart from the six of us (including Petrov) knew anything about it.

# *Breakfast in Azimut*

The next morning, we met in the Hotel Azimut reception before we were all due to leave. We'd had a continental style breakfast in the restaurant, which I was amused to see had an extensive range of pickled produce as well as cheeses, hams and salmon.

Christina and Irina both saw a couple of people in the corner of the restaurant acting suspiciously and moved in on them. They turned out to be backpackers trying to get a free breakfast on the hotel!

Then, we took a cab back to the airport and Amy and I braced ourselves for the nearly whole day combination of flights to get us back to Bodø. Amy said she could probably have used a Brant jet, but we both thought it was better to travel without trumpeting from the highest points where we were going.

Christina and Irina were lucky. From Moscow, Christina could fly directly to London and Irina even had a direct two-hour Aeroflot flight from Murmansk back to Saint Petersburg.

We said our goodbyes and split up for our various flights. Amy and I would see several airports before we returned to Bodø. Amy and I talked about our findings now, and wondered what the next Jump would tell us more. It was Jump 20 - which amounted to a whole year. Strictly speaking it was 364.09 days and I could tell that Amy was tempted to come along again. Now that she could see the unfolding mystery, she wanted to see where it led.

"You know something? I'll ask Hermann to sell my car, and I'll come along again for the next Jump. It's only one year. I'll need a good story for the Lab though, for all of us actually."

"It's easy," I said, "We can become part of an advisory group to the Chinese. Say we are going to work at SuzGene for some time. It will be enough to deflect suspicions and I don't think Brant really cares enough about our project ever since everything went wrong in Barcelona."

## *Aport snap-fit cartridge*

The next morning, we were all in the Lab. I'd greeted Juliette the previous evening, and we decided, based upon this absence, that it must be intolerable to be the ones waiting for a Jump to finish.

Work had started on our new second lab already, and the requests for us to travel to China were being handled via Hermann. He'd also got a good supply of each of our mugshots for visas, passports and permissions and we'd all left him several copies of our signatures, which he could drop into official applications.

Next, we climbed into the Typhoon. Amy, Juliette, Rolf and me. Hermann powered on the accelerator and Rolf ran the countdown for the Jump from inside the Typhoon.

"3-2-1."

Then there was a noise at the door. Hermann opened it. I looked for signs that we'd made the jump but couldn't see any.

"Come on out," said Hermann, "And welcome to the

future." I realised Hermann was wearing a different shirt. We had made the Jump 20 – a whole year.

"This time I made a list of the changes!" said Hermann.

1.  Your 'Trip to China' and personnel swap was approved.
2.  Amy's car sold for a good price. It was still factory fresh.
3.  Building of the new lab completed - now we have a decoy.
4.  The lab builders were Russian speakers. I was concerned in case they tried to steal anything.
5.  Dr Irina Sholokhova made contact, but I said you were 'away' and she understood.
6.  The incoming SuzeGen people have designed some good nano-technology equipment, but we think they stole the design from Biotree.

"There's more to talk of over a beer, of course," said Hermann, "But I think these are the headline items. And - it's good to see all of you again!"

"Did anything happen about the meteors?" asked Amy, "For example, did they start signalling?"

"No, although there is a prediction that Earth will encounter another asteroid belt in around three months' time. No scientists are worried by this and say that it will be a similar non-event."

I walked back into the lab and found myself lost. They had moved the walls around and the whole place was reconfigured. I felt hugely disorientated.

"I'll want you all to troop around the lab complex on a high-speed fly-by, please," said Hermann. "To remove the rumours that I am some kind of serial killer and have disposed of each of you!"

We all laughed, but I could see that the normally carefree Hermann was slightly stressed.

"And we should go to the pub," I said. Hermann and Rolf immediately agreed. It was time for some more of that hyper-expensive Norwegian beer.

On the way to the bar, I called Irina in Moscow. She sounded flustered and asked if she could call me back. We carried on to the Hundholmen Brygghus bar which Hermann had discovered, complete with its built-in microbrewery. When I received the drinks menu, I could completely understand why Hermann liked the place. There were about 100 beers listed including some reaching 10.0 ABV, so very strong.

There was a separate page on their beer menu printed with about a dozen beers which I'd consider drinkable in quantity. I was having a flash back to when I'd first met Hermann and Rolf in Geneva, all that time ago.

Amy and Juliette joined us and the five of us pondered the next Jump. Two years. Amy had said she was out. She said that the idea of the Jump was quite seductive, but she was worried that it she continued any longer then she would never be able to break the habit.

Juliette said she had experienced some of the flashing lights that I'd described, but these had been after the Jump. We worked out that by now Juliette had been on several Typhoon Jumps. I wondered if Lekton was trying to reach her, too.

Rolf said he'd experienced nothing, and Hermann joked

that it was because Rolf was part man, part alcohol.

"But I'm also surprised that we've not heard any more about those meteors that arrived on earth?" said Amy.

"No, the bigger news is about the biomechanical progress being made with the nano-machines, " said Hermann, "Biotree is in the process of designing something called an Aport cartridge, which is a cannula-like device that can be used to administer drugs and vaccines. Now that everyone on Earth is expected to take a vaccine once a year to protect from various illnesses, such as COVID, H1N1 and PCV, to name but three of the common ones, then there is a demand for delivery systems."

Hermann continued, "The latest idea is that some nano-machines can be incorporated into the serum to fix up certain common conditions, as well as to act as a sterile buffer between the serum and the bloodstream."

"I suppose Brant is trying to monetise the whole thing?" asked Amy, "It doesn't surprise me."

"That's right," said Hermann, "There's two parts to the system. The cartridge components and then the snap fit serums. The serum can be configured by region too, and there's even a thought to include different strengths according to socio-demographic groupings."

"What? Like a super-serum for the rich?" I gasped.

I could only think of cameras and lenses and the way that a camera maker would also make a specific mount for their lenses which would prevent say a Sony lens from being attached to a Nikon camera. Proprietary lock-in, I think they call it.

# Game of Chicken

The next few jumps would be like a game of chicken. You know, where you have to stand in the road whilst the truck approaches (not recommended).

I knew the next jump would be for two years, and if I did it with Juliette, then she was the one I cared about the most.

I wasn't sure about Amy, she looked as if she was tempted. She needed to think of the ongoing excuse for being away. For me, and by implication for Juliette, we could simply give up our jobs if we were found out. But for Amy, she was a rising star in Biotree and a three-year gap (i.e., one year plus two years) would certainly be noticed.

Rolf was fully committed to the Jump, and Hermann equally committed to staying behind, so it worked for all of us. Then Hermann had a bright idea.

"It's easy - we can ask for an extension of your loan to SuzGene. And you know what, I think they have already replied!"

Triumphantly, Hermann produced a forged letter from SuzGene. It most graciously requested the extension of both our involvement in Hangzhou and of the people provided to Biotree by SuzGene as a part of the Thousand Talents Plan.

"I've been busy over this last year and I've cleared everything with Astrid Danielsen from Brant HR. You have freedom to act. To stay, or to Jump!"

Hermann continued, "The beauty of this is that the Chinese people working in Biotree will do as they are told. If the letter says 'stay,' then they will stay.

"You travellers will become invisible to the bureaucracy in China. Although they have some of the best surveillance systems in the world, your non-appearance works in our favour. They can't follow up on people that they cannot trace."

Hermann said, "Oh yes, one more thing. I've built another Cyclone while you have all been away. We had the spare one anyway, to use with the second system, but I thought it would be an excellent test to build another complete one. And Rolf, don't worry - I've run it through all the test suites. It is as good as the one we normally use, and I've paired it to the same computer."

Hermann lifted the new Cyclone, with a number 4 written on it. Rolf examined it.

"It looks as if you have done a wonderful job, Hermann!" said Rolf, "In what feels to me to be about one minute!"

"Naja, I thought we could extend the testing protocol," said Hermann, "We could run two Cyclones together, and at short range - maybe Matt and Juliette?"

I looked at Juliette, who was nodding furiously, "It's a great idea, and if we achieve anything new, then it will be great to put it into the news! We need to keep this Lab funded!"

Amy then spoke, "It is agreed then, subject to safety protocols. And Hermann, confirm to me that there is still no direct connection from the Cyclone to the experimenter?"

"Nein, es gibt keine, - I mean there is no connection, the new device is based upon the design of Cyclone 1."

We now had the documentation to be able to extend our time for another two years, which was the length of the next jump. Jump 21. 728 days.

We looked at one another. This had gone well past Miss Bianca and Miss Behavin. Now we would be reaching out to Lekton and future science, in a toroid shaped universe we could transmit a gravity wave across, to awaken a sentient life force that understood us better than we did ourselves.

# PART THREE

## Mad science?

That evening, Juliette and I were talking in my apartment.

"It feels so strange. On the one hand we have been away for a year, and on the other it seems like a matter of days!" said Juliette.

"In my case, even stranger because of the side trip to Russia," I said, "I'm completely disoriented."

Juliette remarked, "Well, the apartments have still been cleaned and I can see that Hermann arranged for the larders to be re-stocked!"

It was true, the apartment still appeared to be almost new, and the refrigerator contained milk, beer, wine, cheese and anything else that Hermann thought necessary to continue civilised living.

"So, what do you think about tomorrow?" asked Juliette, "I mean, you have been on several Cyclone jumps, for me it will be the first?"

"It should go fine," I mused, "Like when I link to Irina in Moscow, only with a much shorter travel distance. You'll have to say 'Hi' to the rat first though!"

"You mean Billie?" said Juliette, "The one we usually use is Billie, and then Tré is the black rat we introduce second."

"I'd forgotten that you are already on first name terms with them both!" I said, "Do you know, when I show up now, the white rat usually sends me a message about 'not you again, no trouble please!' "

"I could see that in Billie's personality. Just wanting to get along."

"How daft is this?" I said, "We are talking about conversing with rodents! Yet we are supposed to be serious-minded scientists!"

"I think the rest of the Biotree Laboratory thinks of us as mad scientists by now," said Juliette.

## Easy-steal wheels

Back in the lab, we were ready for the Cyclone test. Before we had entered the original lab, Hermann had taken us to see the newly constructed lab, which we intended to use as a decoy.

It was incredible! I thought the first one was modern, but this new one was even more stylised. It had clean cupboards, equipment bays, patch panels, everything that you'd want to conduct physics experiments, without looking like something from HP Lovecraft or Bram Stoker.

Hermann pointed to a set of equipment.

"There," Hermann said, "I rack mounted the Exascale in this room and then placed it all into a flight-case enclosure, with wheels. I made it easy for thieves to take away."

"Of course, the outside panels say Exascale, but it is a Dell computer inside; respectable, but not crazy. The wiring and the Cyclone are the originals from our old second system and the software are up-to-date but annoyingly un-patched."

"I've set it up so that it will self-identify its LAN each time it starts, and the cameras on the front will take nice little pictures of the operators."

Hermann had built a perfect trap for anyone deciding to steal the system. It was like those trap-cars they sometimes put into parking lots.

"What about the Typhoon?" asked Amy.

"No one outside of the lab knows about that, and it is also quite difficult to steal a room," said Herman, "Although I thought we could label the washrooms as Typhoon 2."

It relieved me to see that Hermann's sense of humour was undamaged by our absence.

I put on the Cyclone headgear, and Juliette did the same. We were ready for the experiment, although we had to wait for Amy to get the particle accelerator running. It seemed crazy to me we had more-or-less taken over Brant's accelerator.

I had almost forgotten that feeling of being on a precipice before the Cyclone fully kicked in. Then the rat saw me and realised I was online. It gave a deferential message of 'no trouble please,' to me, but then spotted that another greyish rat with a white face (One I had been told was called Nibble, not Tré) was also in the cage. I realised it was Juliette's rat, but my rat was staring towards it, looking blank.

Then I felt the second stage of the Cyclone kick in, I was going on a visit to Lekton. I needed something new from him.

# *Lekton*

Soon enough, Lekton began, "You have brought someone new to me this time. A female. She is also clever, by your human standards. A psychologist, she has maps of you in her mind. Some of your weaknesses, and some of your presumed strengths. There is another one too, someone called Levi. He seems to have many issues, including outbreaks of violence. I don't think she has ever told you about that.

"I can tell that Juliette is filtering her knowledge of you both like some kind of experiment."

"How do you know about Juliette?" I asked.

Lekton continued, although I realised, he was accessing my own thoughts, "Juliette thinks that Levi Spillmann was an almost legendary creature in Israel, He had invented something, seen its downside potential to be used as a military weapon and then disabled it. A rebel with a cause and a philosophy that rejected all absolutes and talked of freedom, authenticity, and tough choices."

I could feel Lekton gently probing around in my brain. It wasn't an unpleasant sensation, and it felt as if I was in the hands of a skilful practitioner.

He continued, "Levi Spillmann's actions and their own garb of troubled sophistication will have been what attracted Juliette toward Spillmann. And the irony is that I - we here - owe him something for his knowledge of how to open the communication portal from humans to all of us. Although I know it was an accident - a side effect- of the design of his system."

I realised Lekton thought the Createl design and the link to the Cyclone were a happy accident.

Lekton continued, "Now it's you and Juliette, as partners in a glossy modern love affair lived out between science labs, cafes and jumps in time - even more spirited because you must keep it secret from many. But Juliette knows it is not as fulfilling because you are less driven than she is."

I realised what he was doing. He was trying to spook me.

Lekton continued, "Juliette doesn't think you have the same spark as Spillmann, but your virtue and interest to her is because you are the test case for your lab experiment. It provides her with a necessary fascination. She is captivated by what you represent more than who you are."

Yes, he was trying to get me riled. I assumed there was a similar process being manipulated on Juliette.

Lekton added, "You fantasise a succession of women, each one meaning everything for a moment like some Satre existentialist. It will play well to the time hops, but with Juliette along there will be limits to what you can achieve."

Now he was trying to sow seeds of distrust.

He added, "I can see that Juliette thinks you are two of a kind, and your relationship would endure as long as you do: but it could not make up entirely for the fleeting riches to be had from encounters with others."

I thought I would try to end this line of discussion, "Is Juliette experiencing this conversation?" I asked.

There was a pause and then Lekton replied, "No. I gave Juliette to Matson, another Persona in these wires. He has his own methods although I can see you two are transparent with one another so will exchange the facts of this encounter - or maybe most of the facts, anyway."

The pause gave him away. I decided he was having to invent some of his suppositions, and I'd managed to wrong-foot him.

He continued, "I can see you have a long relationship of supposed equals, but it will be for time to determine which of you is the stronger of the pairing."

I could interpret this like a horoscope now. Lekton's magic hold was defeated. I could feel his hands slipping away from my brain, as if he could also give too much away with close contact.

Then he added, "But I can see you seek more knowledge. I must provide it so that your experiment can continue Earthside. It is the only way that both I and Matson will ever be able to be released from this, although it still feels too soon in the predicted schedule."

This was good. Lekton was changing the subject and had

moved from trying to out-psyche me to now wanting to tell me things to keep our experiment running. He had passed his ability to ad-lib.

He started again, "Earthside must make some of its biggest mistakes before larger forces prevail. You should expect to see the Earthside fight back against its virus of overpopulation."

"First, with pandemic, then climactic change. Change that suffices to flood the land and boil the ocean. It will last so long that humanity will need to regroup, inside new walled citadels with others trying to scratch a living in wasteland devoid of everything. I predict that some land-masses will be erased from human minds during this period."

Then he added, "Humanity will gather the gifts from the universe and deploy them. Among them will be one particular source of rescue, but there will also be several temptations which could deflect humankind from the path."

I was thinking - Chinese fortune cookies mixed with some wholly realistic sounding predictions. That Swedish schoolgirl had been right about Earth's endgame, if we were to believe Lekton. But then he said something else.

"The muon discovery. Earthside scientists from both Fermi and Brookhaven locked underground, surrounded by rings of magnets and coolant. Testing and analysing for months at a time. A tiny subatomic particle seems to be disobeying the known laws of physics - and really it is."

Lekton continued, "You on Earthside are about to make

one of the fundamental discoveries. It is the first of a chain. It will show you that there are forms of matter and energy vital to the cosmos that are not yet known to human science. Your particle célèbre is the muon, akin to an electron but far heavier, and an integral element of the cosmos. Muons do not behave as predicted when shot through an intense magnetic field because of other discoveries you will make which lead to the evolution of the Trigax - a muon particle projector. It can be used for good or evil. Humanity must decide."

"Trigax," I thought I had already heard that word.

But now, I could feel something else running through my head, as if a separate conversation was also in progress. It became louder and I realised I had also been talking to Juliette. Just as I tuned in, I could see the rat again. It was standing next to the black rat and both were eating chocolate buttons. Then I fell, but I wasn't sure whether it was down or up.

# Back to reality

I was aware of the others Amy, Rolf, Hermann and Juliette all looking at me.

"You were gone a long time," said Hermann, "And you didn't seem to trip any of the safety cut-outs. We had to power down the Cyclone to bring you back. But your heart rate and other systems were all running normally. I think you have acclimatised."

I looked across to Juliette, She was sporting a definitive wet-look. "Um, I guess you were not expecting that!" I asked, mildly amused to see the normally graceful Juliette looking so wrecked. Amy had found a towel and had draped it around Juliette.

"Yes, it was intense. I met Matson, who seemed to know all about me. I could not tell whether it was genuine or if he was using cold-reading techniques, like fortune tellers. You know, shotgunning to hope that something will stick, or using the Forer effect offering vague apparitions to reinforce certainties, like the ruse of saying something positive and then offsetting it with something negative."

It had sparked Juliette's psychology training, at the very least.

She continued, "I could see how they could play me, but some of the information was very specific. He knew of my time in the mountains behind Geneva but didn't play me back a predictable 'Heidi climbs the mountain sequence'. Instead, it was about the shooting accident which I witnessed when I was about seven years old. My rough-neck uncle was out shooting with an ex-war rifle when it suddenly exploded in his hand. It tore his beard away and everyone said he was lucky to be alive and that his beard must have saved him. I'd all but forgotten the event, but Matson accessed it. Then, he started to compare you, Matt, with Levi. I assumed he was trying to drive a wedge between us?"

"Lekton did something similar with me, but I pushed back against it and then he turned to telling us about the future on Earth."

"Yes, the same happened, disaster scenarios and gifts of knowledge from outer space?" said Juliette, looking wide-eyed.

"Yes, that's about the size of it," I said, " I think we had very similar experiences. But I'm not sure what we can tell the Biotree business units?"

"There is something, very personal," said Juliette. She whispered something in my ear.

"If that is true, then we can tell them that two Cyclones linked together can read minds!"

I thought about what Juliette had just whispered. Amy looked towards me expectantly.

"Yes, we can safely tell them that the Cyclone can read minds. Not exactly our original plan, but still a very useful one for some militaristic states!" I said. I wondered what to say about the Trigax.

# Trigax

I cleaned up, after the Cyclone session and then Amy offered to buy me and Juliette a coffee. We were in the canteen. I decided to ask Amy about Trigax. Had I imagined it? I thought I'd heard of it around the labs.

"Well, yes, I have heard about it," said Amy, "It's a piece of Brant Research which Biotree took over. Supposedly defensive, but I'm not so sure." She waved her arm toward the window.

"It's over there," she explained, "In the domes."

She had gestured to a row of small domes which gave the impression of giant golf balls alongside a secondary road in the campus.

Then she continued, "The Trigax was one of the most secretive devices available here - far higher security than our Cyclone. It was intended for atom separation as part of building new nanostructures. It could operate with extreme power and could select individual atoms for isolation. The secret in the mix is the use of muons, which are similar to electrons but much larger. Their breakdown creates the power that the Trigax uses.

I thought about how a device could amplify the signal from a sub-atomic particle, affecting its mass to do so. We were in Einstein territory. I could just about imagine how this could happen with a muon.

"You said it 'was' a super-secret?" Juliette asked.

Amy continued, "Such was the nature of the device that the scientists called it a 'god device'. Although having a local range which they used for experiments and nano-machine building, the boosted calibration could deliver the same power and capability anywhere on the planet and also probably as far as the moon into space. It was simply a matter of getting the coordinates set. I'm told we could drill a hole through the moon if we programmed it to do so. They 'slugged' it so that it couldn't do damage. That's when they declassified it."

I could see how Brant would want to monetise this little death ray project, and all the governments who would line up to make down payments.

Amy continued, "There were various safeguards included that limited the Trigax range to an experimental area designated within the R&D facility and the levels of failsafe were such that it effectively had a huge electronics and software guard to prevent it being aimed anywhere else. As a matter of fact, it is an area that Levi Spillmann worked on. They classed the technology as munitions, and the current peaceful use came with stringent conditions, more like those one would find on a nuclear warhead."

Then she said something consistent with what Lekton had said, "None of us know how the original design was created. It was someone called Holden, who had passed out the theoretical papers to allow the whole device to be built. The quantum science within the device is beyond most folk here."

I thought I would have to ask about Holden, "So where is he? Who has met him?"

"No-one seems to know anything about Holden. He is said to live as a recluse. He just surfaced once to leave the working documents for the Trigax but has not been seen since. Brant immediately wrapped the papers in Non-Disclosures and Patents! They had a sense of the potential!"

I realised that Lekton's predictions about humanity discovering new technology were beginning. It was just that it sounded like alien technology to me.

# *Touch*

The next day we all went to the 'decoy' lab. There was to be a meeting with Morten Lunde, arranged by Astrid Danielsen and including Dr Rita Sahlberg. Amy had told us all to put on tidy-looking clothes and 'to look professional'.

We'd all assembled around the meeting room table in the new lab, when Morten arrived with Astrid and Rita. I briefly saw Astrid throw a glance toward Amy, and it was only then that I realised.

I recollected Amy had been good friends with Astrid when we were back in Geneva, and then I remembered Astrid had been able to secure us all such preferential transfers to Bodø. I was also intrigued that we'd been able to continue with the experiments, including Amy's absence, even when I'd have expected Biotree, or more likely Brant, to get irritated. Of course, it was Amy van der Leiden who was managing all of this, through the key influencer Astrid Danielsen. No wonder! It made me wonder how much about the Typhoon Astrid knew.

Morten began, "Congratulations to you all in this Lab. It is not so often that get to celebrate a major scientific breakthrough of the kind you have achieved with the Cyclone. I've seen Dr Rolf von Westendorf's videos of the last session and seen that with two Cyclones you can operate HCCH (Human to Computer to Computer to Human) Interaction. I'm told that the intermediate computers can also carry additional functions like speech translation, so this has multiple possibilities.

"Of course, we will need to reduce the cables snaking from the Cyclones, maybe with the use of networking? It will make it far more practical to use the devices in operational environments."

"I assume you include 'on the battlefield' in those operational environments?" asked Juliette.

"Well, our ultimate corporate owners, Brant, bought Biotree and their core business relates to battlefield logistic support," answered Morten, "So I guess you are right in your supposition."

Morten continued, "But we must not forget, they make much of the hive mind in instrumenting C3I (Command Control Communications Intelligence) structures. That and the OODA loop (Observe, Orient, Decide, Act). The holy grail has been to augment both processes and to transmit the outcome to a dispersed force. They could be for coercive reasons or simply to contain problems that have already occurred. Even ones in a civil setting."

In other words, make military smarter by enhancing C3I and OODA via Artificial Intelligence.

Juliette's knowledge kicked in for the next piece. Curiously, I found I could predict what she would say. I realised that the last 'mind-meld' must have transferred more of Juliette's knowledge to me, in much the same way that I gained Irina Sholokhova's knowledge on the long link to Moscow.

Juliette began, "The hive mind implies something organic. Evolutionary algorithms (EA), particle swarm optimization (PSO), differential evolution (DE) and their variants. You still need some animal-style intelligence in the system. Matt told me, Morton, that you had once asked him about whether a computer-controlled cockroach was still afraid of fire?"

I realised I had never said this to Juliette, but I was thinking it right now, because my interview was the last time I'd seen Morton and Astrid, together.

Juliette continued, "That's the clue, there are two operating systems. I'll call them the Presence and Persona."

It startled me when I heard this. They were the same terms I had first heard from Lekton, even as far back as our experiments in Geneva.

"Imagine the autonomous operation of the Presence, like the Limbic system in humans. Or more precisely, the regulation of visceral autonomic processes, like breathing and heart rate. Now layer on top of the Presence, like a body onto a chassis, the Persona. What makes this individual look this way? Why is this one a sports car, not a van? They both have the same chassis, but different appearances. That's what we are dealing with here."

Juliette looked slightly surprised that she had just said all of that. I wondered if we were witnessing more of Lekton's new discoveries emerging.

"Swarm intelligence (SI) is the collective behaviour of decentralised, self-organised systems, natural or artificial. They have limits as I will show in a moment. The concept is employed in work on artificial intelligence. SI systems consist typically of a population of simple agents or boids interacting locally with one another and with their environment."

"Here's my example. I look after the various critters we have in the Lab. It is mainly mice and rats, but we do have some smaller insects too. For example, some cockroaches. Now this might gross you out, but one day, we had an accident when a male praying mantis got out of its tank and into the cockroach tank. Males mantis can fly, you know. The cockroaches attempted to attack it, but the mantis climbed up one side of the glass and then started eating the cockroaches. It learned quickly that it was unassailable and could eat as many cockroaches as it liked. The cockroaches, in turn, realised that they could not successfully attack the mantis, even when they mobbed it, but the cockroaches that had been disabled by the mantis became fair game for other cockroaches. It took us about ten minutes to sort everything out, but by then the cockroaches were in a state of carnage."

There you can see two different biological systems operating. In both cases the agents follow very simple rules, and although there is no centralized control structure dictating how individual agents should behave, local, and to a certain degree random, interactions between such agents lead to the emergence of "intelligent" global behaviour, unknown to the individual agents. The cockroaches swarmed the mantis to try to overcome it. The mantis learned it was unassailable, the cockroaches learned about cannibalism of their species."

I added, "So I guess this applies to ant colonies, bee colonies, birds flocking, hawks hunting, animal herding, bacterial growth, fish schooling and microbial intelligence?"

I wondered where some of those examples had come from, until I realised, I was sharing some of Juliette's knowledge. I could see Morton and Astrid looking slightly mind-blown by the last few minutes. Rolf and Hermann gave their own impression of swarm behaviour by nodding intelligently at what Juliette and I had just said.

Amy picked up the conversation, "So Morton is right. The application of swarm principles is called swarm robotics while swarm intelligence refers to the more general set of algorithms. Our original Createl system is used in forecasting problems. Then Selexor can devise synthetic collective intelligence. That is why we must customise parameters when a sufficiently convergent state is achieved and so an optimal solution can be derived."

I thought that Amy was running a snow job over Morton. Telling him he was right and clever, flattering him in acceptance of our position.

Rolf spoke, "It is what we have done for the current experiments. This is no conjuring trick. Cyclone 1 is reading the content of Cyclone 4, and vice versa. We have a two-way brain-wave exchange."

Morten asked, "So what else do you need? I'm going to tell Corporate Brant about our progress, I imagine they will ask us to secure it like we used to secure the Trigax. What is it called? Officially, I mean?"

"We still use RightMind around here, as well as Cyclone, Selexor and Createl," answered Amy.

I noticed she had not mentioned Typhoon. I could see that Astrid looked as if she realised too.

Then, as suddenly as they had appeared, Morton and Astrid left the lab, not even pausing to ask about the equipment. I realised that Morton was like Kjeld Nikolajsen in Geneva, a manager focused on the numbers, living his life one spreadsheet at a time.

# Crisp white wine

That evening, I was in the apartment with Juliette, we had poured a crisp white wine and were sitting on the balcony looking over the view toward the *Soløyvatnet* Lake and into the distance toward the town of Bodø. To the left we could just make out the snow-covered mountains across the *Saltfjorden.*

"So peaceful here," said Juliette, "The way the snow damps the sound."

We idly watched a couple of cross-country skiers make their way past the apartments along the ski-way which had been cut as wide as a road. We hardly noticed it was the other side of a strict security fence guarding the outer perimeter of the Biotree campus.

"Did you get any new information?" I asked Juliette, "You know, from the Cyclone? It told me about your uncle and his rifle accident, after all."

"Not really," said Juliette, "Matson wanted to tell me about the argument you had with Heather before you broke up. I'm guessing Lekton and Matson were both looking for intense scenes? Anyway, Heather appeared to say, *"Va te faire foutre!"* - she must have had a pretty good grasp of French, I think?"

"No, I think she had a good grasp of swearing!" I replied, although I'd only learned what Heather had really said to me when I moved to Geneva.

"Did Matson say anything to you about Holden?" I asked.

"No? Why would he?" answered Juliette, "I know of him but not so much recently. We studied him in Uni. From what I remember he was an Australian - something of a legend - He worked on quantum physics, biophysics, nanoscience, quantum chemistry, mathematical biology, complexity theory, and philosophy I remember we had one of his books as required reading. " 'Quantum Decoherence and the Life Force,' I think it was called. I guess it is relevant to what we are doing now, but it went further into the ideas of self-replicating machines and nano-machines that can replicate DNA. Wild and deep stuff."

"I wasn't sure Holden was an actual person until now!" I said, "I half wondered if it was a buddy of Lekton and Matson!"

"Well, Nakamura Research Industries (NRI) would be interested in your theory. I'm pretty sure that Holden works for NRI, based in Tokyo. As a matter of fact, they say he is quite reclusive and there are no photographs of him. Almost unimaginable in this internet age."

I'd never heard of Holden but decided to pick up my laptop and to have a look. The wiki entry showed a beige picture of a tidy-bearded young man, holding a marker pen by the side of scribbled equations on a white-board. It looked to me like a 1980s stock image from Open

University more than a photo of a real person, and I wondered if it had been added as a placeholder to the article.

The copy read convincingly though:

Bruce Holden, Brisbane, Queensland, Australia is a leading scientist exploring the role of quantum mechanics at molecular scales of relevance to biology.

Holden has dealt both sceptically and sympathetically with viewpoints on quantum questions including the origins of life, quantum theory in biotech, quantum theory and replication, ultrafast quantum dynamics, decoherence of quantum mechanics in bio-structures, modelling using nano-machines, scaling from nano to human-sized, quantum metabolics, achieving evolutionary stability with quantum physics, modelling and replication of DNA with quantum technology, trans memetic intelligence, the geometry of memetic clouds.

"Wow!" I said it out loud, "I can't believe this guy. Surely no-one could handle this much thinking!"

"And the modelling," said Juliette, "It is off the scale."

"Follow the thread," I said, "Using quantum theory to build bio-structures, to have the bio-structures self-replicate, harnessing nano-machines to build the structures, creating intelligence in the life-forms with memetic clouds."

"Do you think this is one of the Leaps?" asked Juliette.

"Damn sure it is!" I said, "But I still can't work out how Holden can provide blueprints for the Trigax to Biotree and all of this other material to Nakamura?"

"It is as if there are two of him?" said Juliette "Maybe there are?"

"Or one quick instance?" I added.

"Do you think you will do it? Asked Juliette, "The next Jump, I mean, for two years?"

"I think so, although I'm not sure what we are trialling now. I mean, we don't have proper tests for each jump, nor predicted outcomes."

"I think we are acting like explorers, trying to work out the next frontier," said Juliette, "Where does this lead? Who are Lekton and Matson? What do they know? Have we stumbled into something? What about the science leap forward? And how are we connecting our minds together? Why two experiments: Cyclone and Typhoon?"

Juliette had given a splurge of reasons to continue. I was the most captivated by the thought of a Giant Leap. Some kind of scientific advances, maybe because of gifts donated from another intelligence. I felt we were close to something. And besides, Juliette had sold her Porsche before the last Jump.

# *Your mind and we belong together*

I'd like to understand just why
I feel like I have been through hell
But you tell me I haven't even started yet
To live here you've got to give more than you get

That I know
But they said it's all right
I'd like to understand today
Then maybe I would know who I was
When I was when it was yesterday

So many people
They just seem to clutter up my mind
And if it's mine throw it away
Throw it again once for my girl friend

You find me behind the door
And all of the far-out faces
From long ago, I can't erase

Your Mind And We Belong Together (edit)
Arthur Lee

## Jump 21

This is where it starts to be a long time," said Hermann.

Amy and Hermann were staying behind and Rolf, Juliette and I were making the next Jump.

It was Jump 21, listed as 728.18 days, or just shy of two years. It didn't affect the three of us, although it would certainly have an effect on those staying behind. They would be wondering and waiting for our return.

Amy had decided to stay behind, after many deliberations. She had considered the jump again, but because she and Astrid Danielsen were close, it was too much for Astrid to bear. In addition, it meant that Amy could fend off the questions from Rita Sahlberg and Morten Lunde. In other words she could 'contain the bosses'.

Amy started the accelerator, and then Rolf fired the Typhoon.

Jump 21 – 2 years.

Then I could hear the door being opened, we assumed

we had jumped forward another two years.

Hermann ushered us from the Typhoon.

"Welcome back! Although there's bad news," he said, "This time our lab was targeted by the Russians. It seems that the last time Juliette and Matt ran the Cyclone, Tektorize was able to run an intercept. They overheard everything and realised that the system they had was inferior to the one here in Bodø."

"How would they do that?" asked Juliette.

Harman again, "We think they set up a recording from their Cyclone; no one wearing it, but they have managed to record the buffered transmission of our interaction. We think it was a lucky spin-off on their part."

Then Amy spoke, "This time they didn't use huge military helicopters, like they did for the theft in Geneva. Instead, they brought a couple of our routine supply trucks up to the gates and entered normally. They had threatened one of the delivery drivers who takes things to the Company Store. He drove a lorry, to deliver a refresh of goods."

Amy continued, "Except instead of boxes of cereal and pallets of fruit, the truck was concealing men who manhandled the Cyclone System from the old lab onto the truck and then calmly drove it out to the commercial cargo handling at Bodø Airport. Then they loaded it into a DHL 757 and flew it from Bodø direct to Moscow."

"But how did they get it out of the lab?" asked Rolf, "Surely the security would stop them?"

"You'd think so, but they had one person on the inside,

who helped them get past the gates. It was completely low-key. I was even working at the time but suspected nothing." answered Hermann.

"What did the big chiefs say?" asked Juliette to Amy.

Hermann continued, "They didn't even seem surprised. Morten said we had to expect more industrial espionage as our systems got better, and Astrid agreed."

Amy added, "Rita Sahlberg was devastated, but I said to her it had happened before, when we were in Geneva.

Hermann added, "But we didn't let on about the system in the other lab. I guess when the Russians find out that they have stolen another dud, they will be pretty angry."

"So Morten and Rita still don't know about the Typhoon and the good version of the Cyclone?" asked Juliette.

"No, only we know about the two labs and their differences," answered Hermann.

# *Irina calls*

There was a buzzing in the lab. Hermann looked over to a corner where he had improvised a cell-phone farm, keeping each of the cellphones of the Typhoon travellers connected to a charging source. It was mine that was buzzing. Someone was trying to make contact = a +7 number.

"Hello...?"

"Hello Matt? Is that you? It's Irina. I guessed you would be back, based upon when your cell-phone went dead. It suddenly woke up again about two days ago. This is the third time I've tried actually; I've been calling at 18-hour intervals for the last three days. You will have heard. Tektorize has stolen the second system from your labs. It looks to me as if it has the same defect as the first one, but at least we can connect two people directly together now, using the two Cyclones."

I realised the Irina was still trying to make the best of the Cyclone without the speed improvement possible with Levi's key.

"Have you heard the theories that are running around in

Moscow?" asked Irina.

"No, " I said, "About what? Can I put you on speaker Irina?" I gestured to the others to gather around.

"Er, who is there, please?" asked Irina.

"We have the people here who work with the Cyclone. Only the people directly involved - you have already met Amy van der Leiden, and I think you know about Juliette Häberli? Others here are Dr Rolf Westendorf and Dr Hermann Schmidt. We are the original team from Geneva, all of whom transferred to Bodø."

"Okay, but you must promise to keep this to yourselves," said Irina, "It relates to Tunguska and the near impact event."

Hermann said, "Yes, the Tunguska event was a massive explosion that occurred near the Tunguska River in Russia. It was like the more recent occurrence when a meteor shot over Chelyabinsk and was famously recorded on car dashboard cameras. Only the Tunguska event was before any of the recording technology was available. I think it was even before the First World War."

"That's right," said Irina, "The 1908 Tunguska event was described as a massive explosion that occurred near the Podkamennaya Tunguska River in Yeniseysk Governorate (now Krasnoyarsk Krai), Russia. The explosion was over the sparsely populated Eastern Siberian Taiga and flattened about 80 million trees over an area of 2,000-square kilometres of forest. Consider that that is around ten times the size of Frankfurt am Main. It was thought to be the air burst of a stony meteoroid about 100 metres in size. That is about 3 times the diameter of the Chelyabinsk one that has been seen on video.

Irina continued, "Tunguska was classified as an impact event, even though no impact crater has been found; the object is thought to have disintegrated at an altitude of 5 to 10 kilometres rather than to have hit the surface of the Earth. That would be the effect of the heating as the meteor crashed through the atmosphere."

Irina lowered her voice, "I think this is an early encounter with extra-terrestrial origins. Look, I don't know whether Matt explained, but when we use the Cyclones simultaneously, some information flows between us."

I spoke up, "Yes, I have told them about this."

Irina said, "I know from your memory exchange that there is a theory about metal rain from the meteor shower we passed through a couple of years ago. More than that, I know that the metallic shards decompose - like rust and that they could leave behind seeds. I have been studying this with Denislav Scekic, a colleague who works in the bioengineering faculty."

I looked at Amy and then the others. I could see that for everyone a large concept had just dropped into place.

Irina continued, "We think that Tunguska was a first attempt to lay down some of those seeds, by an airburst rather than a targeted rain. It is estimated that the shock wave from the Tunguska air burst would have measured 5.0 on the Richter magnitude scale.

Rolf exclaimed, "That's several megatons."

Irina agreed, "Estimates of Tunguska energy have ranged from 3–30 megatons of TNT (that's 13–126 petajoules).

An explosion of that magnitude would be capable of destroying a large metropolitan area. I mentioned Frankfurt earlier. It would have disappeared under such an explosion. Remember the American Second World War nuclear bombs were measured in much smaller kilotons.

"How did we get so careless with science?" asked Juliette, "And what has happened to Science for Good?"

Irina continued, "Well, in Russia, we were also careless with the research. Over 1,000 scholarly papers were published about the Tunguska explosion. But it was not until 2013 that a team of researchers published the results of an analysis of micro-samples from a peat bog near the centre of the affected area, which show fragments that may be of extra-terrestrial origin. They are like the shards that rained down from the recent meteors."

Irina paused, "You see the connection? Something attempts to land on Earth. But it is a hard structure that burns up. It leaves some small shards behind that somehow get through Earth's atmospheric protection. Then a second attempt which uses a metallic rain, whose protective shards get lost in Earth's soil. I think we have received seeds from somewhere unknown."

"What about the older ones? The shards that somehow managed to get through? Is there any sign of either decomposition, or life?" I asked.

Irina continued, "The shards at Tunguska were heated to extreme temperatures and the supposition is that anything inside of them would have been destroyed. Denny (that's Denislav) used his position to get hold of a couple of samples from Tunguska. I can best describe them as around 12 centimetres long, shaped like a one-centimetre thin carrot, with a blackened silver covering. It looks like the covering was melted under the speed of earth entry. We cut into one and the entire inside was blackened and compressed, but with an appearance of glassy carbon or graphene."

"So quite a high-tech container?" asked Rolf.

Irina continued, "That's where Denny was uncertain of what we were looking at, because he thought that the internal material might also have been a housing for a seed. In other words, the shards were high speed transports to get the seeds to earth and their internal structure was of a fullerene mesh - that's like a hollow sphere which could contain payload. Unfortunately, they were no match for the Earth, and the entire internals of the shards were heated and crushed beyond any survivability. And by crash landing in one of the most inhospitable places on earth, it meant our chances for research had always been very limited. Look, I am taking a risk calling you. This VPN link is warning me I am at the end of my anonymous time slot. I must go."

Irina hung up suddenly and without further warning.

"Wow!" said Hermann, "That was pretty mind-blowing. Irina thinks that a couple of earth events might be attempting to lay down seeds. And it goes back to 1908?"

"It is strangely consistent with what I've been told by the voices in the Cyclone," I said, "Earth is sending out a signal and the universe is responding with gifts of knowledge."

"Okay, said Amy, "But we must consider potential downsides too."

A low-pitched siren began in the lab. "What is that?" I asked.

"Earthquake alert," explained Amy, "Nothing to worry about, but they have been getting more frequent around here. I blame climate change."

"And Iceland," added Hermann.

# Vanavara Trading Post

**Testimony of S. Semenov,
as recorded by Russian mineralogist Leonid Kulik's expedition in 1930**

At breakfast time I was sitting by the house at Vanavara Trading Post approximately 65 kilometres south of the explosion, facing north.

I suddenly saw that directly to the north, over Onkoul's Tunguska Road, the sky split in two and fire appeared high and wide over the forest.

The split in the sky grew larger, and the entire northern side was covered with fire. At that moment I became so hot that I couldn't bear it as if my shirt was on fire; from the northern side, where the fire was, came strong heat.

I wanted to tear off my shirt and throw it down, but then the sky closed, and a strong thump sounded, and I was thrown a few metres. I lost my senses for a moment, but then my wife ran out and led me to the house.

After that such noise came, as if rocks were falling or cannons were firing, the Earth shook, and when I was on the ground, I pressed my head down, fearing rocks would smash it. When the sky opened up, hot wind raced between the houses, like from cannons, which left traces in the ground like pathways, and it damaged some crops.

Later we saw that many windows were shattered, and in the barn, a part of the iron lock snapped.

# Arithmetic and Jumps

Sitting in the lab, we realised we had all done the arithmetic for the next series of Jumps. None of it sounded too threatening, until you worked out that from Jump 20 - which had been one year, to Jump 23, which was 7.98 years, a total of 14.96 years would have passed. To us, inside of the Typhoon, it would seem like it had passed in a flash. Literally minutes, yet to anyone outside of the Typhoon system, it would seem like the full 15 years had passed.

So here we were, working clandestinely, in a concealed lab, and only about half-an-hour away from the 15 year forward Jump that was possible. Hermann had been 30 years old when we started. He would be 44 when we reached Jump 23 and then he'd be 60 years old after Jump 24.

By comparison, we'd be about a day older than when we first entered the Typhoon. Now we were playing chicken and the big truck was bearing right down upon us.

# *Time*

*Time, he's waiting in the wings*
*He speaks of senseless things*
*His script is you and me, boy*

*Breaking up is hard, but keeping dark is hateful*
*I had so many dreams*
*I had so many breakthroughs*

*Perhaps you're smiling now*
*Smiling through this darkness*
*But all I had to give was guilt for dreaming*

*David Bowie (adapted)*

## *Are you sure you want to do this?*

"The next jump is for four years," said Hermann, "Are you sure you all want to do this?"

I was certain, and so was Juliette. I was not sure about Rolf, because I could see him thinking about his buddy Hermann and how long he, Rolf would be away.

Amy had declined, much to Astrid's relief. She said something interesting...

"We should sound out Irina for the next jump, although I doubt whether she will want to go."

By now, we had reconfigured the inside of the Typhoon. It contained a modest amount of supplies in a small refrigerator, but we all knew that they would not be needed. Amy started up the particle accelerator. Rolf fired the Typhoon again.

Jump 22 – 4 years.

Once more we had a tap on the door as Hermann opened it. He looked a little different. I could not say exactly, but he somehow looked tired.

"Hey everyone, welcome back!" he uttered.

Well, he was as jolly as usual.

Hermann explained the news: Biotree wanted something back for their investment. They had brought us all from Geneva to Bodø and then loaned us to the Chinese (or so they thought).

The Chinese scientists from SuzGene had devised the proprietary cartridge clip that was needed to hold and release the serum. They had even come up with a branding for it - Tropus. It was being promoted as the next big thing in healthcare, but Hermann had some worries.

Hermann explained, "They have let the Chinese scientists from SuzGene copy the technology and smuggle it back to China. The SuzGene plant in Hangzou now can manufacture cloned product, although they would still need to reverse engineer the Tropus serum and I'm not convinced they know enough to rebuild the nano-machines. It is a very successful theft of proprietary secrets by SuzGene."

"But is the new serum being used yet?" asked Amy.

"No, it is still theoretical rather than practical, " explained Hermann, "But it won't be long before Biotree has combined the new Tropus with the cartridge and will offer timed delivery of medication."

"Now, if you want to keep this programme running, then I think you'll need to run a further Cyclone test and hope to find out something new. The Tektorize experiments don't seem to go any further and so something headline grabbing would be extremely beneficial."

"Tomorrow," said Amy, "You will need time to recover from all these culture shocks first! "

A we had all trooped out. Amy was sitting at a lab table with Astrid. It was the first time we had a 'stranger' present when we emerged from the Typhoon.

It was still such a strange feeling to walk into the Typhoon and then out again years later. I had the strangest feeling of a scene with Juliette in my mind. Juliette and I were on the beach. She was sitting opposite me, eating seafood, but with a faraway look. Amy's words about acclimatisation almost fell on deaf ears. The sensation of the travel took as much acclimatisation as travelling in an elevator. It surprised me that Amy had already forgotten until I remembered it was four years ago. Then I suddenly realised I was losing the context of the present. Who was in government? What was happening around the world? I was losing the Earthside plot.

Although, I still noticed:

Amy was preparing to go on the next Jump with Astrid. For our reappearance, Amy was trying to show Astrid that there were no side effects. The next Jump would be eight years. It was like an Elixir of Youth for them both at this point.

I wondered if Amy had been planning it, and how long Astrid had known. I guess they had worked out the calculations and decided that Jump 23 was the optimal time to do it.

But first we needed news. Hermann asked us all: "Were there any side effects?"

Juliette was the first to reply, "There was some break-through of the flashing lights, but that was all, and it all happened so quickly too!"

I agreed, it was the first time I'd really noticed the lights inside of the Typhoon, like the ones I'd experienced in the Cyclone.

Rolf was shaking his head, "No, it all seemed very effortless. Close the door, blink, open the door and its 4 years later!"

"I could see Amy whispering something to Astrid, but I couldn't tell what. I assumed it was some words of re-assurance. I thought I would brave the question...

"Astrid, are you going to join us for our next experiment?"

Amy looked towards Astrid, who spoke, "Well, since Amy told me about this, I've thought it could be interesting. Like a life booster, almost. If Amy will go through it again, then so will I!"

I knew it. I'd guessed correctly. Only Rolf looked surprised. Our next jump would be eight years.

Rolf looked over to Hermann, "You know something," he said, "My friend Hermann here will need some technical support for the next jump. I'm staying here."

He looked toward Hermann, who smiled. I realised Hermann had missed his drinking buddy. Hermann had always looked slightly older than Rolf when I'd first known them, but now Hermann looked decidedly older than Rolf - almost as if he had prematurely aged. I think

Rolf realised it too, and we both put it down to the no doubt accelerated lifestyle enjoyed by Hermann in the bars around Bodø.

"Wow," said Juliette, "A radical crew change! It'll be me, Matt, Amy and Astrid next time. It is strange though, because we don't have time onboard to bond as a crew, because it feels as if we are only on short elevator ride!"

I knew what Juliette meant. It was the people who we left behind who were really the ones to experience the bonding.

"What about Irina?" I asked Amy, "Did she ever make contact again? I seem to remember you were considering her to travel in the Typhoon?"

"I said there was news to tell, " said Amy, "Some of it is bad. Conditions on earth have deteriorated over the last few years. You'll see it when you go back to the apartments. The earth temperatures have been rising steadily. It has been just a matter of a few degrees. Hardly enough to excite average thinkers, but enough to create special politicised forces in many governments. That happened to Irina, they transferred her to another group, working on climate change in Moscow State University, though still for Tektorize. The Cyclone experimentation ended. But Irina used that special email account to inform us she had received messages on the dual Cyclone from someone called Darnell.

Hermann added, "Globally, we noticed the more extreme hurricane seasons, with floods and loss of life in small areas. It made mainstream news and various aid agencies were dispatched, but it did not really interfere with much of the western world's ways of working."

Then Amy said, "Press reported that relief agencies were corrupted by local country bribes and big business just lobbied to carry on as before. A few large firms bid for the reconstruction work, just as they had done after the middle eastern wars. Brant did fine, because of all the quasi-military reconstruction work."

Hermann again, "The countries closest to the equator are having it the worst. The one-degree rise already creates new droughts and freshwater shortages on top of what was already a dangerously disease-ridden part of the world."

Astrid continued, "In well-developed northern Europe we are aware of what was happening to change economic fortunes, but milder winters and more dependable summers are seen by many as a bonus. You should see how it gets written into the newspapers!"

Amy again, "Now we have the awful blend of climate change and virus attacks, simultaneously destroying the earth. The last pandemic paralleled an early 2000s dystopian movie. Virus riddled bats getting into the human food chain. No-one really wanted to get to the actual cause in case it was engineered."

Hermann added, "Add to that internal combustion powered cars and goods vehicles and their side-effects. Rising economic wealth means the increase in use of cars and air conditioning, but fingers are pointed to methane from cows. Scientists are noticing important changes. The Amazon dropped to a precipitously low level on part of its route. The Mississippi delta became alternately arid and then a major flood plain."

Amy added, "Scientists, like here in Bodø, have been quietly scrutinising the Arctic. The permafrost which had been frozen for thousands of years is thawing. Both poles saw temperature increases faster than the global average."

She paused and then said, "This permafrost dissolved into mud and lakes, consequently destabilising whole areas as the once permafrost-supported ground collapsed beneath buildings, roads and pipelines."

Hermann again, "Think about it...Earlier seasonal snowmelt meant more summer heat went into the air and ground rather than into the act of melting snow, raising temperatures in a feedback loop. More dark shrubs and forest on formerly bleak tundra meant still more heat was absorbed by vegetation."

Amy added, "At sea, the pace was even faster. While snow-covered ice reflects over 80% of the sun's heat, the darker ocean absorbs up to 95% of solar radiation. "

Then Rolf spoke, "Once sea ice melts the process speeds up and becomes self-reinforcing. More ocean surface is revealed, absorbing solar heat, raising temperatures, and making it less likely that ice will re-form next winter."

We all looked at one another. Earth was running towards an endgame. I realised Lekton had been telling me about new discoveries that would help. I'd seen some early signs, like that Australian scientist, and the thought that the rain of shards might somehow bring something to Earth.

## Temperature rise

We agreed to wait a few days before starting another Jump. We caught one of the now all-electric and driverless buses back to the apartment. I looked out of the window. I had expected to come back to winter scenes, like when we had departed. The snow usually lay thick enough for us to need to wrap up warm and wear sturdy boots.

By the time I reached our apartments, I realised that it must be an effect of global warming, like we had talked about in the lab. There was no snow-covered field from my balcony view, the cross-country trails were not visible. Instead, we could see the outer perimeter road, which ran along the outside edge of the fence.

I talked to Juliette again, "Did you get the flashes of breakthrough this time? I did."

"Yes, they were quite intense, like the breakthrough I was getting before I spoke to Matson."

"I always wondered if we were getting some kind of leakage from a data stream. I guess it is Lekton and Matson. Lekton said something once about 'living in the wires'."

Juliette replied, "I've assumed that when Lekton told you that thing about Presence and Persona, it was defining how he operated."

"What do you mean?"

"I think the Lekton and Matson that we experience are both Personas. They are the spooled essences of two individuals, but they would need Presences in order to function like humans."

"You think they are suspended?"

"Yes, I think they are waiting for Earth technology to catch up and then to free them, so that they can transfer back into bodies - or as Lekton described it, Presences."

I realised what Juliette was saying, but also that there would need to be some profound technology transfers to humankind for such systems to be viable.

I looked at the thermometer on the balcony. It read 14C. This was still March and the temperature should fluctuate between minus 2C and plus 3C. Instead, we had almost peak season temperatures, which in a typical July would reach between 11C and 19C. This was three or four months early. Another sign that climate change had begun to accelerate sharply.

Whatever these gifts were from across the universe, they had better arrive soon.

# *Reshuffle*

Back at the apartment, and I was thinking about the next time in the Typhoon.

The next jump would be eight years. It was strangely detached in my thinking now. Around a week earlier, I could think back eight years and it would have some context.

Eight years ago, I was sharing a flat with Tyler Sloan and Kyle Adler. We lived in Kensington and I'd devised a cyber coin mining system. We'd built it from a kit that I'd ordered and then exploited it. It has messed up a few currency exchange mechanisms and ultimately come to the attention of Amanda Miller and Jim Cavendish at MI6 and Grace Fielding at GCHQ. Our coin-mining device was exceptional and had been used to expose currency manipulations driven by Russia and interference from the United States. And we'd been in some scrapes.

After that I'd fallen for Heather and we'd moved to Cork, Ireland where I'd worked briefly for Big Pharma, before being transferred with a girl-friend bust-up to Geneva. That's where we'd tinkered with the original Cyclone, which we realised didn't work as specified and it was also where I saw it stolen twice, once by the Chinese and then stolen from them by the Russians. And I'd met Juliette.

That was all in eight years - except it was now sixteen years ago because of my time travel. Now I was about to fast forward by another eight years.

Would everyone around me have these kind of context changes, or was I unique? Of course, not everyone would be involved with governments and organised crime, but eight years was still enough time for stuff to happen. And I'd lost my current context and it would get worse with another jump.

But my mind was made up, with Juliette I'd do the Jump. I wondered whether anyone else would?

I knew Hermann was adamant that he would not take a ride in the Typhoon. He was still very linked to his family, even if he was separated from his wife. He took regular journeys back to Dusseldorf, to where his family lived, and from what I could make out, everyone seemed to get along.

Rolf's demeanour had changed after few Typhoon trips. He was still enthusiastic about using the Typhoon, but could see his buddy Hermann getting old before his time in Bodø. Simply by being around Rolf was boosting Hermann's mood, so the rest of us let them make their choices.

Now for Amy and Astrid. Astrid seemed to be hooked, like Amy had been on the earlier Jumps. Her view seemed to about preserving youth. Astrid could also help us negotiate a non-interference policy from Biotree.

We'd supplied the plans for the first Cyclones to Biotree and they were working out how to get them into production as well as to gain US Department of State and European Union clearances for the devices. They needed to re-jig all the wiring too, so that it could run as a wireless device, which was no small task. The combination of Hermann and Rolf together could hold the corporate people at bay whilst the processes rumbled very slowly along.

Our challenge would be Dr Rita Sahlberg, who was technically in charge of our lab facility and to whom Amy reported.

"Well," said Amy, " I think we should congratulate Hermann. He has assumed control of the Lab. Astrid helped us to arrange this. Hermann has been promoted for all of his years of loyal service. More money and more responsibility. He will be the one reporting our progress from now on."

I could see the brilliance of this. Hermann didn't want to go in the Typhoon and could be the continuity link amongst us. Amy van der Leiden actually looked to be relieved to lose the responsibility and seemed happy to stay with Astrid Danielsen. Everything was falling into place.

# *Beach*

Days later, were back in the lab, ready for the next Jump. It was going to be for eight years. Hermann and Rolf had refreshed the inner contents of the Typhoon which we reckoned was now already 8 years old. It had still only seemed like about a week to Juliette and I, because we had been inside the Typhon's igloo-like chamber for the time in between.

Amy and Astrid came inside this time. Hermann had given them both pre-flight briefings and checked that everyone was not suffering from any medical conditions.

We clambered into the now familiar seats, and Hermann shut the door. Eight years. A considerable jump.

"Is everyone certain about this?" came Rolf's enquiry on the updated intercom system.

"Yes, we are all ready!" answered Amy.

"Accelerator running," said Hermann.

"Firing the Typhoon!" said Rolf.

Jump 23 – 8 years.

But this time I felt the precipice and saw the flashing lights. I could glimpse Juliette, on the beach again. The same moment. Lekton was trying to make contact.

"The Great Leap is coming," he said, but then disappeared in a shower of lights.

I realised it was Lekton once more trying to communicate, but that our journey time was too short to be in any way useful to him.

Then Hermann was opening the door again.

This time he had grown a beard. He was also wearing round rimmed spectacles, but somehow looked healthier than the last time.

We climbed back out and I could see Rolf had decided to grow a beard too. He looked like a younger brother of Hermann, who had also, it appeared, lost weight.

"Welcome," said Hermann, "Although you are looking less refreshed than normal?" he said.

I looked at the others and saw that everyones' faces looked ashen. This would not be the look that Astrid was expecting.

"What happened?" asked Herman.

"Lekton came. He said 'The Great Leap is coming'," I replied.

"Is that who it was?" asked Amy, "I heard it too."

"Does this normally happen?" asked Astrid, "Only I thought it was supposed to be an easy trip?"

"I can see what has happened," said Juliette, "The Typhoon is creating a gravity wave, which acts as a signal, but this time we must be travelling for long enough that Lekton can reach us. It can still only be milliseconds."

Rolf nodded, "Yes, your trip time that time was actually 4 seconds. It is the longest yet. The last one was 2 seconds and before that only 1 second. I think we are seeing the same multipliers beginning to form, doubling but in milliseconds?"

"It is so strange the way you never seem to age," said Hermann, "Whereas I feel that I'm maybe 16 years older than when you first went inside the Typhoon."

"And to me, it is only about a week ago," I replied.

Amy and Astrid also exited the Typhoon and were both eager to look in a mirror. Of course, they didn't look any different from when they had gone into the Typhoon less than ten minutes ago.

"I don't know...I was expecting somehow to look younger," said Astrid.

Amy answered, "But you do, to all of your contemporaries. You will look the same to yourself in a mirror, because you have stayed the same whilst everyone else has become eight years older. We will both look young around our contemporaries now," explained Amy.

Hermann spoke, "It is fortunate that we moved as far to

the north as Bodø. Things have been going differently further south."

Hermann flicked on a projector.

"I don't remember that?" said Juliette.

Hermann answered, "We had it installed when we had both of the labs modernised. Brant paid. I think they are still interested in this research."

Hermann's projection was showing glacial thaws and wildlife trapped on breakaway floes. It was a pickup from the prior day's conversation, but now with the added context that we'd seen outside of the lab.

Hermann continued, "720,000 square kilometres of supposedly permanent ice has disappeared, and this illustrates the rapidity of planetary change."

Amy added, "That was Earth crossing a tipping point."

Hermann flipped the presentation and it now looked more like a disaster movie on the walls.

Hermann said, "Yes, the mountains are starting to come apart. In the Alps, most ground above 3,000 metres is stabilised by permafrost. The melt zone has climbed right up to 4,600 metres, higher than the summit of the Matterhorn and nearly as high as Mont Blanc. With the permafrost glue of millennia melting away, rocks shower down. These are still early warning signs, yet the summits held by politicians and businessmen in Davos were oblivious to all of this.

Hermann continued, "As temperatures edge upwards, it wasn't just mountaineers who fled. Entire towns and villages were at risk."

Rolf spoke, "But, at the opposite end of the scale, low-lying atoll countries such as the Maldives prepared for extinction as sea levels rose, and mainland coasts – in particular the eastern US and Gulf of Mexico, the Caribbean and Pacific islands and the Bay of Bengal – were hit by stronger and stronger hurricanes as the water warms. Now we've a blend of islands underwater, dead trees and dying coral reefs. None of these things is good for climate balance."

Hermann added, "Another older bell-weather was Hurricane Katrina which hit New Orleans with the combined impacts of earthquake and floods and was a nightmare precursor of what the future held. Most striking was seeing how people behaved once the veneer of civilisation had been torn away. From Katrina, most victims were poor and black and left to fend for themselves by a loftily detached government.

"Four days into that crisis, survivors were packed into the city's Superdome, living next to overflowing toilets and rotting bodies as gangs of young men with guns seized the only food and water available. The USA learnt, after Katrina, to put up tents and use refrigerated trucks to store bodies."

"Why was that?" Juliette asked.

Amy spoke, "Stigma, purely stigma, so that an august building isn't associated with being a morgue, but it was all still cosmetics for the politicians,"

Hermann added, "A memorable scene was a single military helicopter landing for just a few minutes. Its crew flung food parcels and water bottles onto the ground before taking off again as if from a war zone. This was Americans supporting other Americans."

Hermann pressed a button, and a new scene appeared.

"Take a look," he said, "Long lines of people in America. Like the Great Depression. But you know something?" he flipped the screen forward.

A small gasp.

"You may well gasp. These Americans are not queuing for food, nor medical supplies. No. These people are queuing for guns," he announced.

# Time away

"Beyond the climate, what else happened this time while we were away?" I asked.

Hermann answered, "More meteor showers came. They had been forecast and were vectored towards Earth. The scientists said they'd burn out in the upper atmosphere, so I wasn't too worried about it. This was a couple of years ago. We saw the sky-fire as the first meteors streaked past. They all seemed to burn out. It was assumed to be the tail of a vast comet somewhere much further out in the solar system. Early reports made the news and the science community was quite excited. Individual showers didn't make it through the earth's atmosphere. Like most things thrown at earth, they were burned up on entry."

"Then a larger item was observed, on a different trajectory and a course for Earth. This one was different because it seemed to have steering. No-one could work out how it changed course and speed as it approached the earth and it then took a path that allowed it to glide in, to where it landed, which was in the middle of the Woomera rocket testing ranges in Australia."

"Woomera?" asked Juliette, "That's in Southern Australia, isn't it? North of Adelaide?"

"That's right," said Hermann, "It also has the Baker Observatory there."

Rolf added, "The meteorite, or whatever it was, landed further north than Woomera, landing somewhere in the Simpson Desert. They've done fly-overs of the area, but it seemed to bury itself under the surface. It's nothing like Tunguska. The impact from this landing seems to be sub-minimal. More like a plane landing than a massive block exploding from outer space."

"This could be what Lekton was referring to? A gift from the universe?" asked Juliette.

"Possibly, but if it is, then it is very slow-burning. There have been no great scientific leaps - apart from our ones, of course!" said Rolf.

"Unless everyone else is keeping them secret?" suggested Amy.

Rolf spoke again, "There's also been some activity around here. We thought it was seismic to begin with. It is like a large molehill is forming near part of the perimeter. It has formed inside one of Biotree's huge sheds, in Zone A22."

Hermann added, "Yes, we first noticed it about three weeks ago, because it was tripping our sensors for the particle accelerator. Then the seismic team noticed it and put out a report on the Biotree Intranet."

"The pieces add up," said Juliette, "A message from Lekton. Our scientific discoveries, the shards that fell, and then the large disturbance outside Woomera."

Amy spoke, "So what you are saying is that this Typhoon is creating the waves that cross the universe and somewhere in the universe is sending us some improved technologies?"

"And I'm not even sure this is the first time!" added Juliette, "Think about the Tunguska event. An attempt to land something on Earth that exploded in the atmosphere? And the rain of metallic shards?"

"Except the metallic shards have done nothing," Said Hermann.

"We don't think they have, but what if they take some time to establish? Like, say, an oak tree?"

"But an acorn establishes in a few weeks," said Rolf, "At least we can tell it is doing something."

"So we should look for those shards. See if there is any sign of activity," said Amy.

"But where do we start? We don't even know where they landed," said Hermann.

"But we do know that the climate change has continued, unabated," said Rolf, "They call it The Warming now. Between one and two degrees of global warming. The hot European summer has become the annual norm. Even in average years, Europeans are dying of heat stress."

"Can you describe it to us?" Juliette asked.

Rolf continued, "Sure. Agriculture is becoming devastated. Farmers lost billions worth of crops, not to mention extensive forest-fire damage. The flows of the River Po in Italy, Rhine in Germany and Loire in France all shrank to historic lows. The raging fires of Australia and Southern California have spread to Europe.

"Barges ran aground, and there was not enough water for irrigation and hydroelectricity. Melt rates in the Alps, where some glaciers lost 10% of their mass, are not just a record–they are more that double the previous record.

"Crops bake in the fields, and forests die off and burn. The short-term effects may not be the worst"

"From the beech forests of northern Europe to the evergreen oaks of the Mediterranean, plant growth across the whole landmass has slowed and, in some areas, stopped. Instead of absorbing carbon dioxide, the stressed plants are emitting it. Around half a billion tonnes of carbon was added to the atmosphere from European plants, equivalent to a twelfth of global emissions from fossil fuels.

"If these land-based emissions were sustained over long periods, global warming could spiral out of control. It becomes a climate emergency."

"Okay, but we still need to find whether any activity is emanating from the shards."

# *Shards*

Rolf had already prepared a map showing the supposed landing zones for the shards. He had combined the first wave and the second and also added the larger 'landing' near to Woomera in Australia.

"Take a look at this," said Rolf, "We can see that the shards' points of impact seem to be at other notable locations. We've some at Bodø, some at the LIGO to the north of Moscow, there's another landing close to Los Alamos, where the Americans do missile testing, and one in Canada close to Fort Resolution. In China, there seems to be a solitary grouping in Hangzhou, close to the SinoTech research complex. But nothing has ever been discovered at any of the locations apart from a few of the metallic shards, which, when cut into show themselves to be filled with a silicate net structure."

"That, we thought, was the seed suspension," said Juliette.

"In other words, the 'seeds' have been spread fairly widely around the Earth."

Hermann asked, "But what are the new discoveries?...

The Trigax - a weapon. Our Cyclone and Typhoon - which we've kept secret."

Rolf continued, "The last few years have seen a formidable rate of increase in technological progress compared with any period before. We've also got nanotech improvements, the magnalev transit, all the new handhelds, and the beginning of a new drug delivery system called the Tropus.

Some people are calling it 'The Great Leap', because it sometimes seems far-fetched that so many changes happened so quickly. And we don't even have the math to support it - 'it just works', as Apple might say,

Hermann added, "Yes, and there's word of a new pandemic. It appears to be coming from Australia and is called the N3R0 virus. Some people are nicknaming it NERO."

Rolf added, "Brant and Biotree have latched on to the idea, of course. They are trying to complete the design of a delivery system for vaccines, which can be modified. If Juliette was still visible here, I think she would get co-opted onto that team. They are calling the system the Tropus and the Aport is a cannula-like device for inserting cartridges of the Tropus into the bloodstream."

Hermann added, "Brant, with their military contractor links, have drawn up plans which show 'Zones' in a 'managed State'."

Rolf continued, "Hermann is right. Brant says a modern citizen could walk around safely and securely within their designated zones. As long as they have the Aport and Tropus, of course. Then, by a simple application, they could visit other areas, but the system can be regulated to keep crowds and supply and demand for goods and services under control. People with higher status have greater freedom and the ability to travel more widely. They are describing the technology for this as Ellipse."

"These are not incremental ideas, they are leaps," I said, "But are they ideas or does any of it actually work?"

"Oh yes," said Rolf, "It works, or at least some of it does. They have designed the Aport and a first variant of the Tropus. They want to add nanotech to the mix, so that the nana-machines can conduct running repairs to the human body. Some of this nano-engineering is mind-blowing."

"I hope not literally," I said, glancing toward the Cyclone.

"Oh no, and furthermore there are many objections and protests to what is seen as the policing of a totalitarian state," replied Hermann.

"However, the emergency in Australia is seen by some as a possible catalyst to increase usage of the Tropus," said Rolf, "It becomes a humanitarian relief mission."

"So we need to decide whether you are going to do another Jump," asked Hermann, "Now it seems so strange to see you return each time looking like you did all those years ago. The next Jump is for 16 years. I will be over 60 years old, but you will all still be the same age! And I feel certain that things will have changed beyond recognition in that time."

I knew I was in. I could see that Juliette was too. I could not tell about Amy and Astrid. We should find out the next day.

## Discernible Pause

We were all assembled outside of the Typhoon. Hermann asked who was going. Juliette and I stepped forward. Juliette hugged Hermann.

"Thanks for doing this, and looking out for us," she said.

Then Amy and Astrid stepped forward, "Astrid is very keen to do one more jump," said Amy, "And I'll want to accompany her."

I could sense that Amy's attraction to the Jumps had now declined, and that it was Astrid driving the pair of them. However, I sensed it would be their last Jump. The arithmetic was getting away from us all now.

We sat inside of the Typhoon. Hermann ran the accelerator and then Rolf fired the Typhoon. Sixteen years.

Jump 24 – 16 years.

This time there was a discernible pause, like when I first used the Cyclone. It felt as if I'd gone over the edge of a precipice, lights and then that memory of Juliette sitting at a table on the beach, eating seafood. Then the voice of Lekton.

"I didn't think you would make it all the way to here. Some of you humans are stronger than you appear. You have passed the point now. The Great Leap has happened, and Earthside will change. Maybe I will meet you soon."

Then a crash that seemed to be in my subconscious, like sometimes when one awakes because of a loud noise, but can't quite recall the noise that caused it.

There was a scraping at the door and then it opened. The young face of a man looked in.

"Hello," I said, "Where is Hermann?"

"Oh, they are both here," answered the man, "My name is Lars Fjelstad. I am following them both with these experiments. They said I could have the privilege to open the door."

We clambered out. Sure enough, Hermann and Rolf were sitting in chairs peering towards the Typhoon. I noticed Hermann was wearing what looked like a hearing aid. The new guy, Lars, was with a woman.

"Hey, everyone!" said Hermann, clean-shaven and noticeably more jovial than he had been for the last couple of Jump tests. I spotted a new attachment to his left arm, which I surmised was one of the Aport injectors for the Tropus. Rolf was similarly kitted out.

"I can see you have already spotted out adaptations to these advanced times," said Rolf, also smiling.

"Yes, the Tropus seemed to be controversial when we last talked about it, " I asked.

"Well, that was before the quarantine restrictions came down for the southern hemisphere," said Hermann.

"The N3R0 virus was highly contagious and not very pretty if one caught it. Somehow the invention of the tropus was sufficient to inoculate people. The southern part of earth was damaged beyond comprehension and we, Earthside, had to put up some kind of defence to protect the Northern half."

"Are you saying that Australia and New Zealand were destroyed?" I asked.

Lars responded, "There - see- Hermann - I told you - they are real places!"

Hermann looked mystified, "No, I know Matt can be a joker sometimes, this is a trick."

"I should also introduce myself, " said the woman, "My name is Carolin. Lars and I are both Norwegians who have been on the base as our first assignments from Uni."

"Which Uni?" I asked.

"University of Tromsø - that's the The Arctic University of Norway," replied Carolin.

"Uh sorry," I said, "I should have realised."

Lars continued, "We'll, we have been studying what has happened in the southern hemisphere, and the Uni sent us here to Biotree to see whether we could find out some more. We started in the Tropus labs but then we heard about what you have been doing and asked to be moved to the Artificial Intelligence labs. That's when we met Hermann and Rolf."

Carolin continued, "Hermann and Rolf invited us for a drink in town and we agreed. It was great to meet some people from the Lab and to compare stories. The co-incidence was, as they drank a little more, they started to tell us about something that they called The Great Leap."

Lars added, "That is what we had been talking about for some time. How did humanity suddenly get to invent several world-class items in a couple of years? Why were the inventions at the same places as the meteor showers a few years ago? We were suspicious and talked about Australia and the landing at Woomera. Rolf and Hermann had never heard of Australia. We couldn't believe it. It was like they had had it removed from their memory."

Lars continued, "That is when we pieced together the circumstances of the Great Leap. We decided that the Australian rumours of a pandemic called N3R0 - Nero-had been used to speed up the deployment of Tropus. The Tropus contained both a vaccine and some kind of neurological nano-engineering which was experimenting with people's minds. There had been a space programme to create the bracelet around Earth and then to populate the bracelet with so-called charms. Each charm contained a Trigax. A massive tuneable weapon which could obliterate anything it was trained upon. The bracelet and charms were used to police the southern hemisphere, with anything that tried to escape being obliterated. It created the so-called glimmer. Rolf and Hermann are unaware of what happened, not even aware of the existence of Australia."

"How is this even possible?" asked Amy.

"We don't know, but we suspect new science in use by humanity," said Lars, "It is the only way we could explain some of this. For example, a new variety of science is evolving, called 'magnetomics'. It is the study of magnetite and graphite to make new power. Magnetite is a mineral and one of the main iron ores, with its conventional chemical formula $Fe_3O_4$. It seemed to be lowly in terms of any power output yet blend it with graphite plus some muons and it goes super vector creating a braiding map."

"You are 'talking-in-cat' again." said Astrid.

"He means that magnetite plus the field effects can generate a tremendous source of power without excessive fuel consumption. Save the Planet," replied Hermann.

"Speaking of which, " said Rolf, "There has been so much happen: We have passed the two-degree world. Now nobody will take Mediterranean holidays. When temperatures were last between 1 and 2C higher than they were in the 20th Century some 125,000 years ago, sea levels were five or six metres higher too.

Rolf continued, "All this 'lost' water was in the polar ice. The 'tipping point' for Greenland wasn't until average temperatures had risen by 2.7C. Greenland was also warming much faster than the rest of the world at 2.2 times the global average. Despite protests, significant world leaders played the whole thing down."

"What even with predictions and modelling?" I asked.

Hermann grimaced, "Yes, it became fashionable for some

of the awful weak politicians to say they'd had enough from so-called experts, implying rather pompously that they were better than the scientists."

Rolf continued, "It took the situation when Miami was set to flood and disappear, as was most of Manhattan. Central London, despite its river defences, flooded. That's when some of the idiotic blustering political class started to pay attention. It was too late, of course."

"Like thermal runaway?" asked Juliette.

Hermann replied, "Kind of, more like an exothermic reaction, where the heat from one stage accelerates the next stage, that's what happened. Bangkok, Bombay and Shanghai lost most of their area. In all, half of humanity had to move to higher ground. The southern hemisphere was more drastically affected than the north.

Rolf continued, "Not only coastal communities suffered. As mountains lose their glaciers, so people lose their water supplies. The entire Indian subcontinent was fighting for survival. As the glaciers disappeared from all but the highest peaks, their runoff ceased to power the massive rivers that delivered vital freshwater to hundreds of millions. Everywhere, ecosystems unravelled as species either migrated or fell out of sync with each other. You can see how the divisions on Earth were forming. You know, here inside the Arctic Circle we are comparatively lucky. We are not getting the same amount of ice and snow, but it is still quite liveable."

Hermann continued, "But where we are now: beyond two degrees of heat increase, mass starvation becomes an immense problem. Millions, then billions, of people face an increasingly tough battle to survive. To find anything comparable we have to go back to the Pliocene Epoch –

The last epoch of the Tertiary period, 3 million years ago. Although, in geological terms, it's not that long ago. However, there were no continental glaciers in the northern hemisphere and trees grew in the Arctic. Sea levels were 25 metres higher than today. In this kind of heat, the death of the Amazon was as inevitable as the melting of Greenland.

"The warmer seas absorbed less carbon dioxide, leaving more to accumulate in the atmosphere and intensify global warming. On land, matters were even worse. Huge amounts of carbon are stored in the soil, as the half-rotted remains of dead vegetation.

The soil carbon reservoir contains some 1600 gigatonnes, more than double the entire carbon content of the atmosphere. But then as the soil warms, bacteria speed up the breakdown of this stored carbon, releasing it into the atmosphere.

"We are into 'end of the world' territory here," emphasised Hermann.

# Cyclone

I'd made a suggestion to the extended team. It was time to run a further Cyclone test. We'd set up the rig tomorrow. Rolf was happy to do this and for Hermann to help. I decided to call Irina. So much time had passed now, without contact, that I thought I could safely call her on her cell phone number.

*"Привет? Привет? Это действительно ты, Мэтт Николсон. После всего этого времени?Privet? Privet? Eto deystvitel'no ty, Mett Nikolson. Posle vsego etogo vremeni?"*

She switched to English: "Hello? Hello? Is that really you, Matt Nicholson. After all of this time?"

"Hello Irina, yes, it is me."

"I did not expect to hear from you again. You know they have forbidden me from using the Cyclone?" I am working on Climate Change now, although it is becoming a nightmare task."

"I wanted to tell you what we have discovered. I will tell you all of it, so that you know and can add it to your thinking. The Cyclone creates some kind of breakthrough to other voices, with names like Lekton, Matson and Darnell. I usually hear Lekton and he tells me that out experiments have created a gravity wave that is so powerful that the rest of the universe can detect it.

"I knew it!" said Irina, "The power from your other experiment is much greater than from the Cyclone. We can pick it up on the MSU LIGO."

"Well, Lekton went on to say that we would be sent some 'gifts' from across the universe which I think are new kinds of science. Big Science which can produce new devices."

"It makes sense because of the sudden discoveries which came from just after that near miss event."

"You mean the one that landed in Australia?" I asked.

"Australia?" answered Irina, "What is that?"

"Australia, New Zealand, Australasia?" I answered.

"Sorry I don't know where that is. Is it a town near to Bodø?"

I realised that Irina hd the same problem as Rolf and Hermann.

"Have you been given an Aport? And Tropus?" I asked.

"Yes, it makes me sharper and keeps away the pandemic virus," answered Irina.

"So lets list out some of the technological advances," I said, "We've Trigax, The Bracelet, Charms, Tropus, Aport, nano machines, magnalev transit, magnetite engineering, improvements to AI, the Cyclone, The Typhoon, biotrace. It is a pretty amazing list of inventions."

"I agree," said Irina, "And some of them we still don't understand. How they work, I mean."

"Now we can examine what has been happening. As the earth is gradually turning into one vast desert, we have new technologies which could conceivably save out planet. I think that is what Lekton was referring to when he broke through during the Cyclone experiments."

I paused.

"Look Irina, I'm running the Cyclone tomorrow at 10:00 Bodø time. I'll be linked in with my colleague Juliette. Why don't you try to run your Cyclone at the same time? See if you can capture some of the conversation with Lekton? I think it could be very useful for you and your work. We should pool our activities now to try to understand as much as possible.

There was a pause.

"Okay, I'll do it. I am not supposed to go back to that Lab, but I am sure I can make it work. I can ask my husband Petrov Makarovich to help me run the system."

"Petrov? You married him?"

"Yes, we have been married for 20 years now. We have two children - well young adults, both at college."

"Congratulations!" I said, "It only feels like less than two weeks ago since I met you in Murmansk!"

"You sound exactly as I remembered you!" said Irina, "I wish I'd been able to keep up with your speed. I worked out that there was some kind of cryptographic gate inside the Cyclone that made it run slowly. I took the source code to my next assignment, but I could never make it run any faster. Whoever designed it was a genius."

"His name was Levi Spillmann. Sadly, he passed away in a boating accident," I decide to edit the story for Irina. No need to bring in the murder or the other countries playing for high stakes. I realised I'd just done what Juliette did when she first explained about Levi to me.

"You know Tektorize stole a second system?" asked Irina.

"I did know," I replied.

"And it means we can also link two people together, via two Cyclones. Please tell me, is it safe? I will ask Petrov to join me in the next run of the system if it is."

"Yes, it is safe. Although you might learn new things about one another when your brains link. And I guess you will hear from Lekton. You need to get past the fall and the flashing lights if you wish to hear Lekton. He implied you were too slow for his system to work. However, you might hear our contact with him, if the Cyclone still buffers it correctly."

"Matt - Thank you for calling me. I will try to be on the link tomorrow."

The line clicked. I assumed Irina was being careful.

I thought if there was anyone else I should call. I suddenly realised, that like Hermann and Irina, they would have aged across the years that I had remained the same. But if most of them were in their sixties now, it was still possible to reach out. After the next Jump, they would be in their nineties.

I suddenly realised that this was my last chance to tell anyone about what we had discovered. I flicked through my cell phone.

Amanda.

Amanda Miller: I doubted whether she was still at SI6, but I knew she would have useful contacts in UK, Europe and the USA.

I texted her.

"Hi It's Matt. How about a Video call?"

Less than ten seconds later, a reply.

"Matt? You have hidden so well. Congratulations! Yes, I'll call you now from a secured line."

My heart jumped as Amanda called. I looked at her face. She wore the increase in years well. Unlike Hermann, she didn't seem to have grey hair, and she looked happy to see me.

"Wow!" I said, "Amanda - you are looking good!"

"Matt?" answered Amanda, "You don't seem to have changed one bit? You're not some kind of clever AI versions of yourself are you?"

"No, It's really me. Ask me anything and you'll get a straight answer."

"That's enough for me to tell. What is the reason for your call? You know, I didn't expect the world to go like this. It's like we are in an endgame."

"It's about that," I said, "You remember I went to Bodø with the others from Geneva?"

"Yes, and then you disappeared. Not even your friends knew where you had gone. I noticed you phone pop up from time to time, but it would stay on for about a day and then you would disappear again."

"Yes, trust me that what I tell you is true. We used the Cyclone to trial HCCH (Human-Computer-Computer-Human) interaction. It works. We discovered a spin-off effect. This sounds loopy, I know, but we could talk to an unknown being which has predicted the future. We also discovered, by accident, a way to time travel although only in one direction. Into the future. It may be 32 years since we last met, but for me it is only two weeks ago. I have Jumped forward in time many times. Each Jump is double the previous one. The next Jump is 32 years. I wanted to tell you what we have discovered. Do with it as you see fit."

I paused to see whether Amanda was taking it all in. She was certainly still listening.

"The device to take the time Jumps is in Bodø. It is called the Typhoon and only a handful of people know about it. A wider group know about the Cyclone for AI linkages and mind interconnection. Rolf and Hermann - who you met - are both well versed in both technologies. Dr Irina Sholokhova at Moscow State University, working for Teknorize, also knows how to work the Cyclone. What she doesn't know is that we have one version with something called Levi's key at the Bodø Lab. It is the only version that works properly. However, I trust Irina now and would tell her the secret of Levi's key."

I could see Amanda taking a short note. I continued:

"In the breakthrough of voices from the Cyclone, there is a being called Lekton who tells us about a Great Leap where we are given new technology to help Earthside solve its current problems. It has been sent because we sent strong gravity waves across the universe and other life-forms could pinpoint our position."

"Gravity waves?" asked Amanda, "This is becoming like something from science fiction."

I nodded, "Yes, it is what we do. We think the gravity waves triggered the incoming meteor storms which affected earth. We think the storms seeded the earth with something, but we don't know what. It seems to be something that will somehow free Lekton and others. I don't think I know much more, and I realise it must all sound crazy. I will be around until tomorrow but we are are then running the Typhoon again and I will be propelled another 30 years forward.

"I'll be around eighty by then!" said Amanda, "Hey you know that Chuck Manners and I are married! We've a child - a girl - her name is Charlie. Charlie Manners. She is something of a firebrand so keep an eye out for her in the future. And If we're still around then look us up in 32 years!"

"Hey, it is great to talk to you again, Amanda," You are probably the last person from that 'era' that I'll speak to!"

"You don't know," said Amanda, "I've recorded this call, I may play it to Grace and to some of the others."

"Oh, well tell them I care for them all, Bigsy, Jake, Clare, Christina and Chuck, and that despite the 30 years, it still seems to me to about a couple of weeks. Juliette and I have done nearly all the Jumps together, and Amy van der Leiden and Astrid Danielsen have done all the later and longer Jumps. Our next Jump is 32 years, and the one after that would be almost a lifetime. 64 years. After that, we start to become immortal."

"So how do you age?" asked Amanda, "It's not like Dorian Gray, is it?" she said referring to the famous Oscar Wilde story of the painting in the attic.

"No, nor like Faust. There's no pact with anyone. We experience only a few seconds per Jump. It means it is only two weeks or so since our first Jump, and that's around how much we have aged. I still look like I did 32 years ago. By tomorrow, I'll have jumped forward another 32 years, but still look the same as I do today."

Then I asked Amanda..."Amanda, you know where Australia is, don't you?"

A pause.

"Can you spell that for me?"

"No matter. Can I ask if you've been given the vaccine?"

"Yes, I have one of those vaccine delivery systems. The oddly named Aport, which I am supposed to keep topped up with Tropus from a cartridge every month."

Amanda was lucid, yet seemed to have the same gap

about Australia in her memory like the others. I wondered if it was anything to do with the Tropus.

"Matt?"

"Yes,"

"You be careful now."

# PART FOUR

# Crossroads

*I went down to the crossroads*
*Fell down on my knees*
*Down to the crossroads*
*Fell down on my knees*
*Asked the Lord above for mercy*
*"Take me, if you please"*

*I went down to the crossroads*
*Tried to flag a ride*
*Down to the crossroads*
*Tried to flag a ride*
*Nobody seemed to know me*
*Everybody passed me by*

*You can still barrelhouse, baby*
*On the riverside*
*And I'm standing at the crossroads*
*Believe I'm sinking down*

*Crossroads*
*Robert Johnson*

# Group Hug

We were back at the Lab, ready for the next run of the Typhoon. Lars and Carolin were making the preparations, watched over by Hermann and Rolf.

"It's an insurance policy," explained Hermann, "I've been following the math, and this is probably the last Jump I'll be able to watch. I'll be in my 90's when you return, and Rolf will be 78, mainly because he took a couple of the Jumps earlier."

"This really messes with your head," I said.

"Oh trust me, it messes with ours too," said Hermann.

Rolf stepped over, "You might need to learn how to trigger these systems yourself by the time the next jump is ready," he observed, "And look, I wrote a short user guide too."

Juliette said it, "Group Hug."

We did it, thinking that the next return would be a whole 32 years away, during which time much could happen.

We were making the actual Jump some how without further words, although we all knew it was likely to be a permanent goodbye.

Rolf produced some paper and a printed smart code, which he stuck to the wall of the Lab, "Sicher ist sicher!" he said, which I took to be some kind of security statement.

Amy, Astrid, Juliette and I climbed into the Typhoon. Lars and Carolin were busy preparing for the Jump. Hermann and Rolf watched on. I saw Hermann wave, for the first time since we started, as then did Rolf.

We could hear the particle accelerator and then a judder as we teetered over the precipice when the Typhoon kicked in. Flashing lights, That vision of Juliette on the beach and then the voice of Lekton.

"You are a wild one. Coming out of the other side of the Great Leap now. Systems have been put in place. Groundwork has been laid for me and others to assume some level of direct communication with people Earthside. We can start to define zones. The Ellipse and The Tract. There are several key areas where the surviving shards have created new Chrysaora shaped like giant Cnidaria, with domes and long net-like structures."

There was a deep bass judder, and then Lekton spoke again, this time slightly robotically:

"I've been given a backstory, you know. My parents had me when the Nero virus was at its height. They were both immune and so am I. We lived in the country called Australia, in a town called Darwin in the Northern Territories. When the virus was at its height, Australia was the original centre for it and the entire country was quarantined. We'd had the Flames a few years earlier, and much of Australia burnt to the ground. "

"The Flames resulted from climate effects. Wildlife was eradicated over a few years, because of the huge bush fires. The fireys couldn't keep up with it. Relentless, and later repeated in Southern California."

This all sounded scarily plausible, but I had to remember Lekton was a voice, an artificial construct that somehow lived in the wires.

"Altogether it was a modern-day tragedy, a side-effect of global heating. Most of the livestock and about a third of the population were killed. "

Lekton continued, "It made Straya a very dangerous place to live. The old joke was that The Northern Territories were filled with nasty critters. They were all out to kill and eat one another, or any passing humans. Freshies, Salties, jellyfish, sharks, spiders. You name it, they'd kill and eat one another."

I noticed that not only had Lekton been given a backstory, he'd also been given some dialect, as if he was really from Australia. I found it slightly comical.

Lekton's voice continued - still as if he was playing from a script, "The decimation from the Flames left many places with unsafe water, which created some of the contagion. The evolution of the Australian virus ran away from the engineering of its vaccine, causing the tropus to be introduced. Biotree simply couldn't keep up with the variants. People went a burnt black colour when they caught the virus. There was no way to stop it. Their blood didn't run yellow or red, it went through a blue colour and then to black."

Lekton continued, "And despite the death of so-much wildlife, there were still buzzards, vultures and rats which seemed to thrive. The whole landmass was horrific, like something out of a disaster movie. Strangely, only people that were on the landmass remember it. It's like it never happened to everyone else."

"Did people try to leave?" I asked.

"You couldn't go in or out. And I'm not talking about a small landmass here. I'm talking about something the same size as North America."

I saw the flashing lights again. I realised it was the end of the Jump.

Lekton continued, but his voice had lost that robotic edge, as if he was now speaking directly again, "Biotree was trialling their newest versions of tropus and nanobots in the territory, under the control of Makatomi, who works for Holden. Speculation was that they were trying to cover up for something. Quarantine restrictions were boosted. Biotree was helping the effort to instigate the new processes.

"Under Holden's orders, they were implementing a series of geostationary satellites, nicknamed 'The Bracelet' which applied an electronic border around the landmass. It was like the End of Days and the politicians and leaders opted for discretion to avoid a world panic. It was thought better to contain the massive problem than to let panic set in globally."

The world juddered and I was back in the Typhoon. Thirty-two years into the future. I was surprised when

Lekton's narration continued:

"I don't have long to say this. Holden will locate me. He put pressure through Makatomi, onto Biotree to come up with a resolution. Desperation led to the Bracelet to which they added the so-called 'charms' which provided enforcement."

Lekton's voice was fading, "Only Tract dwellers and anyone still in Australia know these things. The charms operated with a railgun. It was a way around space-wars legislation because the railgun is not technically classed as a weapon. There's no explosive warhead, yet they fire a high-speed projectile which can cut through anything, meaning enforcement of the territorial edges of Australia was via the Bracelet and charms. Typically, a breach would be spotted, triangulated and then three railgun cannons would be deployed to stop the escape. There was even a Biotree branding for the technology: Trigax."

Lekton finally stopped. I felt a weight lift from inside my head. I realised Lekton had been probing my brain all the time I was hearing the story.

There was a crash and the door of the Typhoon opened. "Strange," said Rolf, "The door stayed locked when you returned. Welcome back."

Rolf looked old. He still had dark hair, now flecked with grey, and was once more bearded, like Hermann had been in the now distant past.

"Hermann?" I asked.

"The old man is here!" said Rolf, "He has been waiting for you all!"

I looked around, and Hermann was in an electric wheelchair, with some kind of monitor. He smiled across at us. He was looking his age now, ninety-something. His eyes looked bright, and I knew he would utter some kind of riposte. I was relieved when he did. Hermann's spirit was still in the house.

"I leave it to the young ones, and they jam the door!" he said, "Welcome back to the Museum!"

A man walked across. I suddenly realised it was Lars Fjelstad, and with him was Carolin. They had both aged well, and looked tanned, fit and healthy. I decided it must be the improved climate affecting Northern Norway.

"Hello again," said Lars, "Hermann is right you know. This part of Biotree was turned into a museum a few years ago. It is cited as the area where the original AI experiments were conducted which led to several key Earthside discoveries. Hermann had the museum named after Amy and Rolf. It is called the van der Leiden Museet, and the individual labs are called the Amy wing and the Rolf wing. You remember that Hermann was put in charge? When it came to preserving the systems, he and Rolf hit upon the idea to create a museum, to stop the lab from being dismantled."

Carolin continued, "We knew there was much more to the lab than most people knew, and so we stayed to ensure it was properly managed. We have been using the Cyclone and discovered the link to Russia. But the most interesting thing is how we can ride our Cyclone Personas onto various Presences which we have created. It allows us to build cloned cyborgs and to layer on to them different personalities. This has given the work here a huge boost."

"It's what Lekton and Matson predicted," said Juliette.

Lars added, "Biotree is now so large that it is considering space exploration to help Earthside get back on to an even footing again. We have heard voices when we are in the Cyclone, and they are telling us about a new form of energy generation - one that can help Earthside reverse its terrible future projection."

"What has been happening here on Earth?" asked Amy, "And why is everyone calling it Earthside?"

Lars began, "Earthside spread as a name, after we worked out that some of the new discoveries could have come from an alien source. And the state of Earth itself? A three-degree increase in global temperature threw the carbon cycle into reverse. Instead of absorbing carbon dioxide, vegetation and soils released it.

"So much carbon poured into the atmosphere that it pumped up atmospheric concentrations by 250 parts per million boosting global warming by another 1.5C.

"All soils were affected by the rising heat, but none as badly as the Amazon's.

Carolin added," 'Catastrophe' is almost too small a word for losing the rainforest. Its seven million square kilometres produced 10% of the world's entire photosynthetic output from plants. Drought and heat crippled it and then fire finished it off. Farming and food production tipped into decline. Salt water crept up the stricken rivers, poisoning ground water. Higher temperatures meant greater evaporation, further drying out vegetation and soils, and causing huge losses from reservoirs. The TV news shows featuring droughts became increasingly common.

Lars added, "Grain yields declined by 10% for every degree of heat above 30C, and at 40C there was no more grain. The Indian subcontinent was choking on dust.

Carolin added, "All of human history shows that, given the choice between starving in-situ and moving, people move. Pakistan was one of the early failed states as civil administration collapsed and armed gangs seized what little food was left. As the land burned, so the sea kept rising.

Lars said, "New York flooded and the eastern part of England also. There was also mass migration away from the stricken areas. It made the earlier populist debates about migrants seem ridiculous, although it allowed resurgent fascist parties to still win votes by promising to keep foreigners out."

Carolin again, "Where they survived, coastal cities became fortified islands. This wasn't pretty though. The world economy was in tatters. A few fat cats bet on the decline of businesses and made huge sums from shorting the markets. They picked high land to develop into prestigious dwellings, with castle-like fortifications and walls around them.

"Direct losses, social instability and insurance pay-outs cascaded through the entire system, with funds to support displaced people increasingly scarce. Certain of the politicians were also dipping into the money to be made from the catastrophe. Building works, Infrastructure, military and medical aid, not to mention shorting the equities in weakened companies. It was cynical, amoral feeding from the trough."

"Earthside cannot deal with the rate of change. The poles are melting, projecting a 50-metre rise in sea level. It could take millennia to complete, but even the metre every 20 years that did occur is way too much for civilisation to handle.

"China was also on a collision course with the planet. As its people became richer and could consume at a rate similar to America, they were eating two-thirds of the entire global harvest and could burn through 100m barrels of oil a day, or 125% of the world's output.

"It is worse because China's agricultural production also crashed, and it was left with the task of feeding 1.5 billion much richer people on two thirds of current supplies. That's why Sino-Nihon is being discussed. The joining of China, Japan and several other smaller countries–frankly to avoid expensive conflict in the region.

"Now, they have formed an Earth Council to try to make sense of what was happening. A problem is that right from the start it contained many vested interests. These were people from major corporations who saw the angle to try to gain control of a larger slice of the planet.

"They are presiding over the Earth as we now know it. An entirely different planet. Ice sheets have vanished from both poles; rainforests have burnt up and turned to desert; the dry and lifeless Alps resemble the High Atlas; rising seas are scouring deep into continental interiors.

"One temptation may be to shift populations from dry areas to the newly thawed regions of the far north, in Canada and Siberia. Even there summers may be too hot for crops to be grown away from the coasts; and there is no guarantee that northern governments will admit southern refugees.

"Right now, Siberia is only one stop from war, with Sino-Nihon ready to invade Siberia and let's not forget that America has already incorporated Canada.

"That's when the Zonal Laws were first proposed, and the power of the Earth Council extended. It should have been by country voting, but was declared a global emergency, so that everywhere could be incorporated.

"Of course, it met with huge resistance from some areas. Countries that had been power brokers, or countries that were doing okay, despite everything. It tipped into the Klima Wars, which started diplomatically as a phoney war, but then toppled into an actual war.

"If it had not been for the new discoveries of the Great Leap, it would be the closest the planet has come to ending up a dead and desolate rock in space.

"But these last effects have not happened. Instead, we can reverse the trends through new off-world discoveries. The climactic change can be reversed. We've found new technology that removes the need for carbon fossil fuels. There's a new form of lightweight, yet formidable material too, but we don't yet have the wherewithal to mine it."

"There's a power play too," said Lars. "It is something that I'm quite worried about."

Carolin agreed, "Biotree are moving the management team from our key locations and replacing them with new people. Two of the longer-term bosses here were Mayer and Nikolai, but now both have been replaced, and their entire record seems to have disappeared from the system."

I realised I didn't have any useful frames of reference now. My network was effectively ended.

Amanda might still be around, but she would be a similar age to Rolf. She might still be connected, but she would would have far less of a network that previously. Similarly for the people from Triangle. Everyone would be around sixty years older. I didn't have any move to make.

It's when I remembered what Amanda had told me. About her daughter with Chuck Manners. Charlie, I think she had said. A firebrand. It sounded promising.

# Running out of road

Juliette was sitting with me in a hotel room in Bodø. Hermann had cancelled our apartments years ago. He'd shifted our funds into some fancy equities and then let them ride. It turned out to be a good move. Everything had multiplied beyond our wildest expectations, including my fund from the old Coin days. Juliette and I were, as they say, 'Minted', as were Amy and Astrid with a combination of equity funds and now vested share options in Brant.

We had to decide what we'd do next. The time travel illustrated to us our loss of currency as we'd now moved away from almost everyone we knew, most of whom were in the twilight of their lives.

"I'm going to try Amanda again," I said. I dialled her number and was surprised to get an almost immediate pickup.

"Hello Matt, I kept your number in my address book, and I thought you'd be back around now."

Amanda flicked on the video feed and I could see her: now grey-haired, but still looking slim and very much engaged in the conversation.

"Chuck is here, and says 'Hi'! " She moved the camera on her phone to pick out Chuck, looking remarkably agile for a seventy-something year old.

"Your next Jump will be out of our lifetimes," said Chuck, "You'll be another 64 years on."

"If we decide to do the jump," I said, thinking it was my first hesitation, "And hey - you remember Juliette!"

"Still hanging in there, Juliette?" said Chuck, smiling.

"We've been hearing about Earthside," Juliette said, "And the challenges whilst humanity tries to make sense of everything."

"We've talked about what you both said, several times," said Amanda, "Both with Chuck, but also with Grace and the Triangle team."

"I think there's been several forces at work," I said, " There's been Lekton, predicting that a series of great discoveries would occur to help Earth survive the current situation. He has also spoken about how he, and others could become freed to be able to act. Then there have been the various corporate machinations. I can see that Brant and Biotree are trying to make financial gains from the discoveries."

"And how!" said Chuck, "They are one of the few mega-corporations on the planet now. And about to start a space programme. They have such a large run-rate of revenue from the tropus."

Amanda said, "At least they have had, but now there's a

Chinese corporation called SinoTech cloning some of Biotree's technology. You should check with your people in the lab though, because I think BioTree is still streets ahead in the manufacture of nano-machines."

Amanda added, "And it wouldn't be a call to me if we couldn't give you some intrigue too. We think that there will be a plan to divide Earthside into three segments. One containing America and Canada, another containing Europe and Russia and the third containing China and Japan."

"What about everywhere else?" I asked.

Amanda explained, "There will be well-developed areas contained inside of something called the Ellipse. Outside of it are rougher areas called the Tract. Some parts of the old Earth are no longer habitable at all because of the Scourge, the Flames and the Klima Wars. To be honest, most people have already forgotten about those areas."

"Like Australia?" I asked once again. Both Amanda and Chuck stared blankly back.

Then Amanda said, "But hey, I told you about our daughter, she is here at the moment. Charlie, say hello!"

The strangest thing then, as I saw someone who looked like Amanda when I first knew her, appeared in the video stream. She didn't dress like Amanda though. Far more 'street' and she moved in the athletic way I remembered Christina moving.

"Hi Charlie," I said, "I hoped we could say hello, I've known your parents for decades!"

"Oh," said Charlie, "You are that time traveller scientist?

- I've been working with AI and nano-bots and stuff like that too - you are way younger than I expected," she added.

"I've only been travelling for less than three weeks, during which I've travelled 60 years," I explained.

"I heard about it from a warez forum," said Charlie, "We heard about a Russian woman - a Doctor - who was trying to do the same thing. And then several others in Norway succeeding? Very cool."

"You know what the speculation is?" asked Charlie.

"No, tell me?" I asked.

"We think that there's some kind of nano-machines which can alter perception. That's how the Southern hemisphere is being blocked out. Biotree is somehow shipping nano-machines within the bloodstream as a result of their tropus vaccines. If it's not that, then there is something else running neural controls to capture and modify knowledge. It is outside of known science."

Charlie's grasp of science seemed to be instinctive rather than through scientific endeavour.

"I'm just off to meet my friend Scrive Mallinson. He's another hacker-adventurer like me. We're both freelance and we both somehow make our living operating on the edges. I've just finished tinkering with a casino - a bit like your own crypto-mining, maybe?"

"Maybe, "I answered, "I was more interested in manufacturing the currency, rather than winning it!"

Charlie continued, "As it happens, Scrive just been

contacted by someone from Biotree. Their name is Makatomi. They want Scrive and I to do something."

"Makatomi is the big boss here."

"I know, I'm thinking that this must be something big, for Makatomi to contact Mallinson."

"You should stay in touch," I said, "You can ask your parents about me if you like!"

"Oh, they've already told me about you. Matt the cyber-coin scientist time-traveller. It sounded bonkers until I've actually met you. But I'm not sure I'd want to Jump forward another 64 years! You'll have to find a way to dial it back. Look, I'll send you my details. And its good to meet you! And you, Juliette! Both of you stick around in this time-zone!"

Charlie left the call.

Amanda smiled, "Well now you've met my whirlwind daughter. There was a time when I'd worry about her going to London to meet a strange freelancer. Scrive she calls him - everyone is getting unusual names nowadays. Grace has been married to Zeno for years. Is it from Greek philosophy or that Dragon Ball game? I don't know!"

She added, "But watch out, my professional instincts still tell me there might be trouble from organised crime in all of this. Most organised crime are operating on a strategic level, offering access to their resources and networks. These groups offer everything from Afghan heroin and Russian methamphetamines, to cybercriminal expertise, and investment of dirty money. More pointedly, they have been able to invest in big business as a way to hide their criminal cash flow."

I realised that some of the thematic crime from my days with the cyber money were still playing out.

Amanda continued, "Russian organised crime invested dirty money into struggling European football clubs to launder proceeds from crimes at home and to run and rig illegal betting operations in Portugal, Austria, Estonia, Germany, Latvia, Moldova, and the United Kingdom. Don't you hate soccer's corruption?"

I was aware of the corruption in sport and the particular greed around the cash flow of soccer, which was largely unseen by the fans. Only the time that the FIFA rigging emerged was there any real focus, but then it slipped away like some underground serpent.

Amanda continued, "Nowadays Organised Crime use a SuperCorp to straddle their territories and drop their ill-gotten cash flow in alongside all of the legitimate money. A Russian president popularised the method many years ago, having learned it on the streets of Saint Petersburg. I'm sure that is how Biotree's blended model is working. The concept of 'Russian-based organised crime' (RBOC) can be a most useful term for capturing the essence of this relationship between criminal groups and the state.

Amanda added, "It is being copied now by the Chinese, and even the Americans are looking for their slice of the action. But know that if someone gets control of the head of this particular serpent, then there will be direct consequences for the world."

We carried on chatting for a while. I could see that Amanda and Chuck were as sharp as ever, but they had both lost some of their links into their sources for intelligence. It seemed that Grace Browning was similarly deprived. Out of GCHQ, with a head of memories but no direct access to any systems.

At one point, Amanda said, "You know something, Matt? You've given me an idea! It's not an entirely original one either!"

She didn't speak of it further. We'd reached the end of the useful line. Any more Jumps and we'd be launching ourselves into unknown territory. Juliette seems to be of the same opinion. With no sort of network and a limited frame of reference, we would be easy pickings.

Juliette said it, "Tomorrow we should go back to the lab and talk things through with the others. We need to understand what they have found out, too."

# The Museum

The next morning, we all sat together in the lab. I had wondered how Hermann and Rolf were even still employed by Biotree, but now, from a small plaque attached to the wall I could see that they had both been awarded the pinnacle Gunnar Ragnvald Honoured Fellowship of Scientist Awards, which gave them a free ability to roam the labs.

Juliette, Amy, Astrid and I were decidedly younger-looking than the others. Lars and Carolin joined us but seemed weirdly older than us now, despite being younger when we first met them.

Hermann had invited Dr Rita Sahlberg to the session. Rita had continued to run the lab for many years and at some point, Rolf and Hermann had let her into the secret of what we were doing.

Rolf explained, "Yes, our lab had credibility because the HCCH linkages between humans lead to an inevitable increase in interest from Brant, who wanted to monetise everything."

Rita added, "Yes, I pushed through the programme to get

the Cyclone's complex wiring remapped into a wireless system and so the new generation Cyclones could be worn without needing cables."

Hermann added, "There was a downside, which you'd feared, Juliette. Of course, Brant wanted to monetise Biotree's product for potential security use."

Rolf said, "We all knew what that meant. They wanted to militarise it."

Hermann continued, "We told Rita about the Typhoon just after the discoveries of the Aport. Biotree had so many proprietary inventions on hand by that time, that the thought of another - the Typhoon - would be too much to handle."

Rita nodded, "Yes, its a case of news management. Only so many stories can be processed simultaneously. Wide time and all that."

"Oh yes," I thought, "Wide time - what politicians instinctively understand. How many simultaneous narratives can one person process?"

Rita explained, "The main Bodø lab was researching nanobot enclosures to supplement the Aport's ability to handle measured vaccine release. It was a smart money-making scheme. If everyone on the planet needed vaccine to counter the pandemic virus, then it was a limitless licence to generate cash, derived from the vaccine sales."

She continued, "This is about when we first heard of Holden, as a new boss of Biotree. He installed a new operations person - Makatomi - who took a first-hand control of how we were operating. It is fortunate that we had already created the Museum by that point. Makatomi followed the spirituality of Buddhism and was loath to de-venerate the Museum which had been set up based upon the teachings of Rolf and Amy. He saw it as Theravāda (Pali: "The School of the Elders") and Mahāyāna (Sanskrit: "The Great Vehicle"), although, ironically, he did not know about the great vehicle of the Typhoon."

Rolf continued, "Makatomi even had the entrance of the museum area remodelled, with Buddha's Four Noble Truths. He said the goal of Buddhism was to overcome suffering (duḥkha) caused by desire and ignorance of reality's true nature, including impermanence (anicca) and the non-existence of the self (anattā)."

Hermann spoke, "There was a fundamental downside through all of this though. The Particle Accelerator has been decommissioned. It was deemed to be too expensive to keep running. If only they knew."

Rolf added, "Yes, but then Makatomi commissioned the new Brant and Raven-funded launch site called Sakhalin-III on the island of Sakhalin to the north of Japan."

"Sakhalin?" asked Amy, "Isn't that a major Russian oil-field?"

"It was, but it got shut down during the Klima Wars," answered Rolf, "The island was disputed territory between Japan and Russia, with the Russians as the last to invade it. Ironically, much of the oil exploration on there was conducted by Japanese subcontractors, and then it got folded back into Greater Japan, or Greater Nihon, as we say nowadays."

"This is so weird," said Juliette, "It is as if Makatomi is building his constructs around what we are doing, without actually knowing of its existence."

"Cat science?" I asked.

Juliette and Hermann laughed.

# *Holden*

Still in the Lab, I was now pondering what we had heard about, "Lekton said something about Holden on the last Jump. I couldn't quite understand it,"

Juliette nodded, "Yes, I heard it too. Something about: 'I don't have long to say this. Holden will locate me.' - It made Holden sound as if he was chasing Lekton?"

Amy and Astrid both nodded now, and Amy spoke: "Yes, we heard as well. We were talking about that yesterday evening. The strangest thing is, that Astrid can remember someone named Holden in the organisation from 45 years ago. We wondered if this could be Holden's son?"

"I'm guessing that it is more likely to be Holden himself. The same person, at the same age, but 45 years later?" said Juliette, "Like he is on the same path as us."

"But what about Makatomi?" I asked, "How does Makatomi tie in?"

"My guess?" Juliette said, "That Holden has sought out a human to enact his will, probably in return for wealth and power?"

"Doch!" said Rolf, "Here comes Goethe's Faust. You remember: Faust is bored and depressed with his life as a scholar. He tries to take his own life but then calls the Devil for further powers with which to indulge all the pleasure and knowledge of the world."

"I guess you studied that in your German school?" asked Amy.

"Correct," said Hermann, "So Mephistopheles, the Devil's representative, makes a bargain with Faust: Mephistopheles will serve Faust with his magic powers for a set number of years, but at the end of the term, the Devil will claim Faust's soul, and Faust will be eternally enslaved."

Rolf spoke, "He's editing the story for us!"

Hermann continued, "During the term of this devilish bargain, Faust makes use of Mephistopheles in various ways. In Goethe's drama, Mephistopheles helps Faust seduce Margarite - (Gretchen), whose life is ultimately destroyed when she gives birth to Faust's son."

Rolf spoke again, "Another piece of editing licence!"

Hermann smiled, "Realising this unholy act she drowns the child and is held for murder. However, Gretchen's innocence saves her and she enters Heaven after her penitent beheading."

Rolf added, "In Goethe's rendition, Faust is saved by God via his constant striving—in combination with Gretchen's eternal pleadings."

Hermann added, "Yes, but in the Polish versions, Faust is irrevocably corrupted and believes his sins cannot be forgiven; when the contract ends, the Devil carries him off to Hell."

Rita asked, "Not bad! I can tell that Faust is taught to all Germans! But think; we don't know anything about Holden, just that he exists and influences everything. Shiraki Makatomi San we do know about.

Rita was looking at her smartphone, "Makatomi, joined Biotree five years ago and has climbed quickly through the corporate ranks holding numerous positions including: Business Intelligence, Human resources, Global Sales and Marketing, Corporate Strategy and Medicinal Safety Research. He was promoted to President and COO last year."

"Was that leveraged on his relationship with Holden?" asked Juliette.

"Undoubtedly," answered Rolf, "He had all the moves at just the right moment."

Rita continued, "We know that Makatomi joined Biotree immediately after graduating from the University of Tokyo, one of Japan's most elite academic institutions, where he graduated from the Faculty of Bio-engineering."

She added, "Then, his last promotion came just a day after Biotree announced a collaboration with the American government, to produce Cyclone helmets to support law enforcement. The collaboration comes amid Biotree's shift towards its partner organisation Brant Industries and their use of military contracts as a primary source of revenue."

"It is like Makatomi has been placed there to work the system," said Hermann.

"And probably controlled by Holden," added Rolf.

That's the moment when I realised that Lekton's predictions were happening.

- Lekton's accomplice, Holden was running things via a human named Makatomi.
- The meteor shards had somehow given Earthside its knowledge to turn itself around.
- Biotree had exploited the tropus to provide vaccine for the entire planet, for huge profits.
- Nanomachines were developing but still didn't work.
- There was some new scientific potential involving magnetite and muons. We just didn't understand it yet.
- Amanda had warned me that organised crime would look for a SuperCorp to run its business. Biotree had become just such an enterprise.

"It's no good," I said, "I'm going to have to use the Cyclone again, to try to find out more."

# Wire-free

We decided to use the new wire-free Cyclones. Both Juliette and I would trial them, although Lars and Carolin had both used them and Rolf and Hermann vouched for their integrity.

If we were running the Cyclones without the particle accelerator, I wondered what would happen. It would be like the earliest days of our trials, although the systems would perform at full speed because of Levi's key.

My interest to do this test was less about any ability to link with a rodent, nor even the ability to link with Juliette, but far more to try to find out more from Lekton.

Juliette had admitted to me the same. We 'd also noted that even the rats now had wireless linkages. We both wondered what Miss Bianca would have made of it and realised that she would be at least 20 rat generations ago.

Rolf and even Hermann busied themselves with the preparations. I knew they were making more of a song and dance of it because Rita was watching. I happily complied with their various calls leading up to the start of the experiment, although I honestly don't think we'd ever run through the sequence in the past.

Then, the old sensation of falling, followed by a shower of lights. They seemed somehow different this time, it was as if the Cyclone had got a better grip on my brain. We had the flashing lights and then things cleared. I'd gone past the rat point without even noticing. I seemed to be looking at the experiment through someone else's eyes. I realised it was Lekton.

"Yes, I'm inside your brain this time," said Lekton, "But don't worry, I need you to continue to function autonomously, so this is just an exhibition of my power. I am what Earthside regard as an Occupier - someone who can occupy another's Presence. Remember when I explained to you about Presence and Persona?""

"Why can't I push back?" I asked.

"You don't have that accelerator. It means we are more evenly matched. But I can't understand why you would come back by this method? Not when you have one that is greatly more powerful."

"We wanted to see you. To ask some more questions."

"About what? I have told you everything?"

"Maybe, but it was probably too fast for me to process."

"What? With that particle accelerator? You were the one that was very fast."

"No. I want to know why you are here? And who is Holden, and what is your goal?"

"I thought you had realised. We are like you. Travelling

through time, but on much greater Jumps. Long enough to have a presence on the Inside as well."

"You mean you are working in the seconds on the Inside of the Jump?"

"Yes, although the seconds double as well. Every time. We are soon travelling huge time jumps, but they give us time to exist on the Inside. Green, Matson, Holden, Darnell, Cardinal and me, of course. All in transit through time. But it becomes a time that has little meaning when the Jumps are as great as ours. So, we grab at the walls. We hold on to something passing. A spike, which gives us a chance to do something in these Inside moments."

Lekton paused, then continued, "I could see you, a fledging traveller with incomplete knowledge. I grabbed at you and those who immediately followed you."

Lekton continued, "Holden is different. Holden has found a powerful motivation. One that will provide him with access to the refreshed world. To the new Earthside."

"In our existence on the Inside, we can travel along the timeline, back to when we entered it and to a long way forward when it might have developed further."

"We can see that Earth will divide. That there will be three Zones, each run by a different politic and under the control of a different enterprise. The catch is that the human discovery to make it all possible has still not been made. All of the other discoveries are ready though. Holden can play a powerful hand now, through Makatomi."

"Then Green and Matson and later me, Darnell and Cardinal. I still have a long time to wait. Probably around 300 Earth years."

"What about us?" I asked Lekton.

"No, you are not for the Inside. You and Juliette are too pure for this life. You will be bounced back to the Outside and never be able to re-enter this world. The same if your friends attempt it. I don't recommend it. Your brains will start to fry if you push this any further."

"What is the discovery?" I asked.

"It's about muonic-magnetomics. The energy source and science to save the Earth. Muon neutrino energy release. Its discovery is due around now."

Lights, judder and I'm thrown out of the system. I'm drenched in sweat again - a factor I'd completely forgotten and then I see Juliette writhing in her chair before being ejected back into the lab.

"Wow, that differed greatly from Typhoon travel. And it felt far more intense than in the old experiments," she observed.

Rita looked at us both, "I've never seen anything like that before in all of our experiments," she said, "What happened?"

We explained that this was not our first time and as we unloaded more of the story, we could both see her expression becoming more incredulous.

"Hermann, you knew about all of this?" she asked.

"Yes, but I don't think it would change anything, although now we know about the Inside, which seems to be a very long wave form of transport.

"Can we summarise? "I asked. I was thinking back to Clare's techniques.

"Yes, " said Rita.

- "Several entities seem to be trapped on the Inside, and are trying to exploit the little hooks they can see in the timeline.
- Holden seems to have found a bigger hook and is trying to use it.
- There is a gap in discovery- something called muonic-magnetomics.
- Everything seems to be paused waiting for another big event."

Rita leaned toward us, "Muonic-magnetomics is another completely secret Biotree discovery," she said, "It is why they built the new cyclotron to replace the particle accelerator. Since they harnessed the muon, tiny amounts of certain materials can produce controlled energy, like a nuclear reaction, just as powerful, but far less dangerous. Think of muons like heavy electrons - something like 200 times·the size, but still leptons. In other words, the most basic of particles with no sub-structure. The thinking goes that muonic-magnetomics can pack an energy punch and are easier to manipulate than, say, electrons."

Then Rita smiled and said, "And most people just say 'magnetomics'!"

She added, "The reason that Biotree and Brant have both

started space exploration is because they are looking for a source of muonic-magnetite. But it will take years to build a mining infrastructure."

"It's the missing piece," said Juliette, referring to Lekton's description. "Lekton said there was something missing. That we'd have to wait to find it. This must surely be the thing?"

Amy asked, "But what is Makatomi doing then, under Holden's instruction towards obtaining the new discovery? And why are Charlie Miller and Scrive Mallinson involved?"

Hermann replied, "Here is a kind of answer. When Sakhalin-II oil and Liquid Natural Gas was decommissioned, Raven and Brant built Sakhalin-III. Instead of digging for energy, they built a space vehicle launch pad, away from prying eyes. Then, 30 years ago, they launched Vanavera-I to search in deep space. It passed Jupiter and Vanavera-I scanned its moons. On Ganymede, they found a huge source of muonic-magnetite. Raven Corp is now involved in the race to mine it."

"Raven?" I asked, "Aren't they somehow connected with Brant?"

"Yes," answered Amy, "It is a complex business structure which reads like a Russian doll because Raven owns Brant which, in turn, owns Biotree. Rest assured that if Raven is involved, then this must be big."

Rita spoke again, "Yes, and it is why they now want access to the secrets inside the Cyclone. They intend to create autonomous beings which can be used to operate the mining on Ganymede."

"Cybernetics? This will be hundreds of years away!" said Juliette.

"I know," said Rita, "or two or three Jumps for someone who wants to see the outcome."

## Liar Paradox and Reality Distortion

We were back in the remaining apartments.

Hermann and Rolf had let most of them go, except their own.

They had moved into Astrid and Amy's apartments, which seemed logical as both Rolf and Hermann were now senior and well-established people in Biotree.

Hermann had Amy's apartment modified because of the internal stairs. He had them exchanged for a small glass elevator between the two floors. Hermann and Rolf had also kept the apartments modernised as time had progressed. There was a new kitchen and revised furniture in both of their apartments now.

Hermann needed the electric wheelchair to get around now, although Rolf was as sprightly as ever.

Rolf had suggested that we all meet together to review our next move.

I'd pulled together a summary list, the way that Clare Crafts used to do, and we looked through it.

- Typhoon time Jumps. The next, if it were ever possible, would be 64 years.
- There is no further ability to start the particle accelerator.
- The particle accelerator created gravity waves which traversed universe and created a signal.
- Distant life forms have gifted Earth new knowledge.
- The Cyclone is now wireless and gives access to Lekton.
- RightMind, even with its many defects, gives access to other's brains.
- Earth is approaching a climactic end-state, but Lekton indicates it is reversible.
- Great Leap of donated knowledge has included:
  - Transportation (Magnalev and Space travel)
  - Energy (Magnatomics)
  - Health management (Aport, Tropus, Nano-engineering)
  - Planetary Management (Ellipse, Tract)
  - Defence (Trigax, Bracelet, Charms)
  - Workforce (RightMind, Cyclone, Robotics, Swarm-Mind, Tracking)
- The shards and other landings on earth had seeded something, but we could not tell what.

Amy added, "Then there were the Lekton predictions that in 300 years we will know how to exploit magnetomics.?"

Juliette added, "By implication, we can see that these technologies build to allow us to get to Ganymede, to mine the magnetite. Many of the other discoveries seem to fit with the space travel necessity. Long distance travel for health-managed humans, supported by cyborgs. Control of workforce via RightMind, Tracking and Robotics."

"And alongside, we have Holden and possibly others, such as Matson to contend with?" asked Amy.

"And how do they exist?" I asked

"In the wires?" said Juliette, "As some kind of constructs?"

"It is making me think about the Presence and Persona again," I said.

"Like the operation and personality of a being?" asked Juliette.

Rolf looked at Juliette, "You are describing AI-based cybernetics, aren't you?"

Juliette nodded, "Yes. I wanted to make sure I had understood it correctly. Lekton must expect a cybernetic solution for the transportation of magnetomic energy to Earthside."

"We have a slow build, with Holden and Makatomi creating the first steps on a long journey which Lekton expects to reach his frame of existence in around 300 years?" suggested Amy.

Astrid looked on through all of this. She knew, in this room of scientists she might not be able to add much.

Then she said, "Is the timeline fixed? How does Lekton

know what is coming, unless he/it can also Jump backwards?"

"It's the Liar Paradox," said Juliette, "Do we believe Lekton? I know that he was saying some things to Matt to make him aggravated at me, and Matson was saying similar inflammatory things to me, aimed at Matt."

Hermann smiled, "Oh yes, I remember the Liar Paradox!"

Hermann mapped out the examples:

**'Single Liar':**
  (1) sentence (1) is false

or the two-sentence version:

**'Double Liar':**
  (1) sentence (2) is false.
  (2) sentence (1) is true"

Hermann added, "These examples were popular at CERN in Geneva when defining various forms of 'truth'!"

Juliette perked up at his: "Yes. Quantum mechanical cognitive entities! Some deep math with this. I expect Amy can solve it though!"

Amy replied, "But do you see they can also apply to Lekton? Or Lekton and Matson? Lekton tells Matt 'the truth' but mixes in some lies. Or Matson tells Juliette 'the truth' but blends in some lies?"

"We won't be able to solve this," said Juliette, "We should just look for the simple version. No unusual events to distort reality."

Just then my phone rang.

"Hello?"

"Hi, it's Charlie Manners. I'm been working for Biotree, in the nano-engineering labs. I thought you should know."

Now there's a reality distortion.

# End of it all

Three whole, real-time months later and yes, you guessed, it.

Juliette and I got married.

We had the tiniest wedding on Bodø's snow-backed beach, with the laboratory gang along, and a seafood supper. Lekton will surely have detected it.

Charlie had further news about Makatomi before she had to go back to London to meet with Scrive. I think they are involved with something quite dangerous.

We told Brant about the particle accelerator and they recommissioned it. Amy supervised, but it still took a year to get it running again, during which time we got to know Bodø.

Brant was all over the invention of the Typhoon, looking for ways to monetise it. Instead, as a spin-off, they found better ways to use the particle accelerator to help build the nano-machinery.

A nano-engineering scientist named Sheri became a good

friend, and she said her Lab needed to extend the range of nano-constructor parts to build intelligent machines. The machines were so tiny that a misplaced speck of dust was like throwing a planet at some of them.

Sheri explained to us all about the fragility of the nano-machines. Apart from atomic forces that would blast nanostructures apart, the continued jittering from Brownian motion and the protein fuel consumption of the tiny devices, Sheri explained, there were still some basic components that were proving impossible to construct. It was like attempting to build a car, but with only a few degrees of steering and no gears.

But we all knew that the tiny machines would eventually be developed and work, in a Jump or two. And they would support new Artificially Intelligent life-forms to help rescue Earthside.

Biotree should have been going from strength to strength, but we knew that there was an imperfect cloning of Biotree's products from somewhere in China, and it was ripping into Brant and Biotree's share price.

Lars and Carolin were concerned because some of the received wisdom from Lekton was proving faulty. Amy's speculation that Lekton was blending falsehoods into his predictions was accurate.

We could also see that Holden's agenda was impure and Brant had become a toolkit for Holden's aims. Amanda Miller was right about the level of organised crime permeating everything.

One day we all assembled in the lab.

Juliette and I were around 33 years old, and yet now

Hermann and Rolf were around their nineties.

A final group hug.

Then the 'old men' Hermann and Rolf started the equipment for their last time and launched Juliette, me, Amy and Astrid forward another 64 years.

See you in the wires.

Ed Adams

Jump

Ed Adams

Jump

CPSIA information can be obtained
at www.ICGtesting.com
Printed in the USA
LVHW082052180521
687738LV00015BA/1076